Crop Protection Handbook – Grass and Clover Swards

Lincolnshire College of Agriculture and Horticulture, Riseholme

INSTRUCTIONS TO BORROWERS

Please replace on the correct shelf any books you
take down for use in the library room.

Books taken out

Enter your name and date on the card, and place the card in
the box provided.

You may keep the book for a fortnight, after which it should
be returned

Return of borrowed books

These should be returned to the Enquiry Office or placed in
the slot in the library and NOT replaced on the shelves.

Crop Protection Handbook – Grass and Clover Swards

R. D. WILLIAMS

BCPC Publications

2A KIDDERMINSTER ROAD, CROYDON CR0 2UE, ENGLAND

©1984 by The British Crop Protection Council
2A Kidderminster Road, Croydon, England

British Library Cataloguing in Publication Data
Crop protection handbook – grasses and clover swards
   1. Grasses – Diseases and pests – Great Britain
   2. Pesticides – Great Britain
   I. Williams, R. D. II. Haggar, R. J.
6332′08995′0941

FB 608.G8
ISBN 0 901436 70 4

Printed in Great Britain by
The Lavenham Press Limited
Lavenham, Suffolk

CONTENTS

Contributors   vii

Foreword   viii

Editor's Preface   ix

List of Tables   xi

List of Figures   xi

List of Plates   xi

Chapters

1. The significance of weeds, pests and diseases in grass and clover crops   1
    Weeds   1
    Pests   1
    Diseases   2
    References   5

2. Integrated control of weeds, insects and pathogens   7
    Advances in cultural, mechanical chemical and biological control   8
    Need for compatibility of control techniques   11
    Interactions between weeds and other pests   12
    Additive effects of control techniques   14
    Conclusions   15
    References   15

3. The establishment phase   17
    Controlling weeds during establishment   18
    Pests of establishing swards   26
    Seedling and establishment diseases   28
    References   29

4. The management of established grassland   31
    The soil   31
    Soil physical conditions   33
    Defoliation   33
    White clover   35
    Putting together a system   35
    Further reading   36

5. Controlling weeds in permanent swards   37
    Cultural control   37
    Chemical control   37
    Occurrence and control of common weeds of established swards   39
    References   47

6. Pests of established grasses and legumes   49
    References   52

7. Diseases of established swards 53
   Diseases of grasses 53
     Diseases caused by fungi 53
     Diseases caused by bacteria 67
     Diseases caused by viruses 68
   Diseases of herbage legumes 72
     Diseases caused by fungi 72
     Diseases caused by viruses 75
     Diseases caused by mycoplasma 78
     References 79

Appendix I Identification of broad-leaved weeds and crop and weed grasses 81

Appendix II Safety precautions 88

Index 89

CONTRIBUTORS

R. J. HAGGAR
: AFRC Weed Research Organization

R. O. CLEMENTS
: Grassland Research Institute

A. J. H. CARR
: Welsh Plant Breeding Station

S. PEEL
: Grassland Research Institute, Permanent Grassland Division

FOREWORD

This is an important book about a very important crop. It is the only book that I have seen that deals specifically with the protection of the crop from weeds, pests and diseases and that attempts to assess the importance of these in the context of grassland management and good husbandry.

It is a book about grass and clover crops that are grown to produce a sward of pasture for ruminant animals to graze, or to be cut and conserved as feed. In the establishment phase, such crops can be treated much like any other arable crop and the chemical and cultural methods of crop protection are similar. Thus, Chapter 3 deals with The Establishment Stage. However, once the crop is established and the sward has been formed, all the other complications that result from the way in which it is utilized by grazing animals, and the interaction of the animals and the sward have to be taken into consideration. In order to accomplish this, it is rarely necessary or possible to attempt to obtain maximum production from the crop. In fact, most grassland management systems used in the UK make no attempt to maximize the cropping potential of grass and clover swards. Thus, the effects of weeds, pests and diseases in reducing yields, unless they are in epidemic proportions, are not fully appreciated or rarely considered in the formulation of animal production systems based on grass and clover swards. Hopefully this book will help to put these "pests" in perspective in relation to all the other factors that grassland farmers, advisers and research workers have to consider when devising such systems.

The British Crop Protection Council has as its main objective the promotion of knowledge and understanding of crop protection, and its members are all concerned with the encouragement of the science and practice of weed, pest and disease control. Thus most of its publications deal with the use of herbicides, insecticides and fungicides, and three of them, *The Pesticide Manual, The Weed Control Handbook* and the *Pest and Disease Control Handbook* have become standard references in the pesticide world.

However, recommendations for the use of herbicides in all crops grown in the UK, which was the objective of the Weed Control Handbook, soon became too numerous and complex to include in a single volume. Instead it was decided that it would be more manageable and useful to produce handbooks that dealt with specific crops or groups of related crops, and to write about crop protection as an integrated approach to the use of pesticide chemicals in combination with all other appropriate techniques (cultural, biological). A series of books was therefore planned that dealt, in the first place, with cereals, sugar beet, potatoes and grassland.

It was soon discovered that writing such books depended very much on the ability of the Editor to weld together the specialist approaches of his authors into a holistic approach, and was much more difficult to achieve than writing handbooks devoted solely to recommendations for the use of pesticide chemicals. It is to the credit, therefore, of Dr. R. D. Williams and his authors that they have been the first to produce a book in this series.

Much of the work of the British Crop Protection Council is undertaken on a voluntary basis and this book is no exception. The authors are all specialists and authorities in their own fields, and all have full time employment in Institutes of the Agricultural and Food Research Council. We are very grateful to them and to the Editor for producing a book that should do much to stimulate interest in an aspect of grassland farming which has so far received too little attention.

E. K. WOODFORD
Director AFRC Weed Research Organization 1960–64
Director The Grassland Research Institute 1964–76

EDITOR'S PREFACE

The British Crop Protection Council has previously published the *Weed Control Handbook* (3,4) and the *Pest and Disease Control Handbook* (5), both dealing with the principles and practices involved and covering the vast array of organisms adversely affecting all farm crops in the UK. However, it was felt that a need existed for a series of suitable handbooks centred on individual crops. This particular handbook is intended for use primarily by farmers and advisers concerned with control of weeds, pests and diseases in grassland.

The treatment of weeds, pests and diseases will be seen to be rather uneven, in that there are positive and specific recommendations for chemical treatment of most of the important weeds, but not for the pests or diseases, although control measures are described for the two latter where they exist. The reason for this is that weeds are the most obvious and easily recognizable of the organisms concerned, so that more research and development have gone into controlling them. It is only recently that the significance of pests and diseases in grassland crops has begun to be recognized, while the economics of chemical control measures are often still in doubt. The point is made, however, that chemical control should be a last resort; good husbandry should keep swards in good condition and only when stress conditions result in an infestation or infection should it be necessary to resort to chemical methods.

The amount of space devoted to diseases may also at first appear disproportionate. However, not only are many of the diseases – particularly those caused by viruses – impossible to identify with certainty except by the use of sophisticated techniques involving expensive equipment and great expertise, but also the pathogenic organisms are microscopic and not easy for anyone other than a trained pathologist to identify. Hence, the descriptions given of them are fairly detailed, unlike the more familiar weeds. It is important, before taking action against diseases, to have diagnoses confirmed by a plant pathologist of the Agricultural Development and Advisory Service or elsewhere. The pests dealt with come somewhere between the two, many of them being small but few microscopic, and many also (e.g. slugs) being easily recognizable and all too familiar!

Chemical names are those used in *The Pesticide Manual* (7) and recommendations (Chapters 3 and 5) are numbered and italicized, as in the *Weed Control Handbook*. Recommendations are given only for approved products and no proprietary names are given; the latter, together with those of the manufacturers, can be found in the *List of Approved Products* (1). Although no firm recommendations are given for control of pests, all the products are approved; a few of those claimed to control diseases are not yet approved, although most of them are.

It cannot be too strongly emphasized that these products should be used only in accordance with the manufacturers' directions and all the safety precautions observed (see also Appendix II). These often include restriction of grazing for a time after application.

Botanical names are those in the *Flora Europaea* (6) and common names of plants those in *English Names of Wild Flowers* (2).

Finally, I would like to express my thanks to Dr E. K. Woodford for his help and advice, to him and the authors for their patience and forbearance and finally, but by no means least, to Mrs E. A. Dean, who cheerfully and expertly typed the final copy for the printer and corrected many of my mistakes.

R. D. Williams
Editor

References

1. Anon. (1984) *List of approved products and their uses for farmers and growers.* Ministry of Agriculture, Fisheries and Food. London: H.M.S.O.
2. Dony, J. G.; Rob, C. M.; Perring, F. H. (1974). *English names of wild flowers.* London: Butterworth.
3. Fryer, J. D.; Makepeace, R. J. (Eds.) (1978). *Weed control handbook. Volume II Recommendations.* Oxford: Blackwell Scientific Publications for British Crop Protection Council.
4. Roberts, H. A. (Ed.) (1982). *Weed control handbook: principles.* Oxford: Blackwell Scientific Publications for British Crop Protection Council.
5. Scopes, N.; Ledieu, M. (Eds.) (1983). *Pest and disease control handbook.* 2nd Edn. Croydon: British Crop Protection Council.
6. Tutin, T. G., *et al.,* (5 Vols. 1964–80). *Flora europaea.* Cambridge University Press.
7. Worthing, C. R.; Walker, S. B. (Eds.) (1983). *The pesticide manual.* 7th Edn. Croydon: British Crop Protection Council.

LIST OF TABLES

Table 2.1   Summary of non-chemical methods of controlling weeds, pests and diseases.

Table 3.1   Amounts of herbicides tolerated by perennial ryegrass at the 3-leaf stage.

Table 3.2   Herbicides for killing broad-leaved weeds (3–4 leaf stage) in newly sown (a) grass/clover, or (b) all-grass leys.

Table 3.3   Susceptibility to herbicides of broad-leaved weed seedlings in newly sown grass/clover leys.

Table 3.4   Herbicides for controlling certain specific weeds in newly sown leys (including those undersown in cereals).

Table 3.5   Wild-oat herbicides for use on cereals undersown with grasses.

Table 5.1   Herbicides for controlling broad-leaved weeds in established swards.

Table 5.2   Susceptibility to herbicides of established broad-leaved weeds in permanent swards.

Table A I.1   Diagnostic features of some common broad-leaved weeds found in newly sown leys.

Table A I.2   Key for identification of vegetative grasses.

LIST OF FIGURES

Fig. 2.1   Interactions of cultural and chemical control agencies.

Fig. 2.2   Interactions in an integrated system of weed, pest and disease control.

Fig. 2.3   A flow diagram of biological and chemical control techniques and their integration for developing pest management systems for thistles.

Fig. 2.4   A diagrammatic presentation indicating the potentials for interactions between different types of pest organisms.

Fig. 3.1   Growth stages of grass seedlings.

Fig. 3.2   Clover seedling at the first trifoliate leaf stage.

Fig. A I.1   Close-up of leaf base of *Elymus (Agropyron) repens*.

Fig. A I.2   Diagram of a grass plant showing diagnostic characters.

Fig. A I.3   Identification of grasses in the vegetative state.

LIST OF PLATES

I–IV   Weed grasses

V   Pests

VI   Diseases of grasses

VII   Diseases of clovers

Chapter 1   # THE SIGNIFICANCE OF WEEDS, PESTS AND DISEASES IN GRASS AND CLOVER CROPS

Despite the fact that some 60% of the enclosed agricultural land in England and Wales is under 'grass' and that over 400 000 ha of grass are sown every year, grass and clover swards are often not given due recognition as crops, largely because they are rarely grown as cash crops, but are used to feed animals which are themselves, or which produce, the crop. However, grasses and clovers are subject to competition from weeds, and to the depredations of pests and diseases, in the same way as are other crop plants.

Weeds

Acutely poisonous plants, such as ragwort (*Senecio jacobaea*), are obviously weeds when eaten by livestock. Chronically poisonous plants, e.g. field horsetail (*Equisetum arvense*) and buttercups (*Ranunculus* spp.), are often unrecognized as such because of the insidiousness of their effect. Other harmful plants can cause injury to livestock, e.g. wall barley (*Hordeum murinum*), or taint in milk, e.g. wild onion (*Allium vineale*). Also there is a whole range of plants which are not eaten freely by livestock, including rushes (*Juncus* spp.) and tufted hair-grass (*Deschampsia caespitosa*); left ungrazed, these species restrict the productivity of the more palatable species.

Not only can these weeds have a directly damaging effect on sward and livestock productivity (25), they often impose added complications and restrictions to field management. For instance, the presence of a few plants of ragwort can rule out silage making from a whole field (14), whilst dense clumps of bracken severely restrict shepherding at key times of the year. The economic impact of weeds in grassland has been the subject of recent studies (13,25).

Most of the broad-leaved and tall growing weeds are obvious to see and identify. However, the same cannot be said of the more prostrate weed grasses, e.g. common couch (*Elymus (Agropyron) repens*) and the unresponsive grasses, e.g. bents (*Agrostis* spp.). Relatively few grazing studies, which permit a direct estimate of the impact of these unproductive grasses on livestock production, have been carried out – except for some notable exceptions (12,15).

The widespread occurrence of weeds in grassland has been highlighted in recent surveys (24); areas infested by specific weeds are given in the relevant sections of Chapter 5.

Reasons why weeds invade grassland are discussed in Chapter 3. Undoubtedly the greatest opportunity for weed ingress is during the grass establishment phase. Of particular concern is contamination by chickweed (*Stellaria media*), especially when it is allowed to form large, smothering clumps – often with infestation as low as 50 plants m^{-2}. Equally, however, annual meadow-grass (*Poa annua*), albeit at high densities, e.g. 2,000 plants m^{-2}, can cause substantial reductions in grass tillering and clover stolon growth, thus limiting the productive life of long-term swards (16).

Pests

The establishment period is also a time when pests can have serious effects; little green material is present and a relatively small number of pest individuals can seriously reduce plant populations. Several invertebrate groups may cause significant damage, the more important known ones being the frit-fly complex (including *Oscinella frit*), leatherjackets (*Tipula* spp.) and slugs (e.g. *Deroceras reticulatus*). Nematodes may be important on sandy soils.

Losses are more widespread and severe than was previously realized (10). Damage occurs in many newly sown fields but goes unnoticed for a number of reasons. For example, many of the pests involved live either beneath the soil surface or within plant tissues and are easily overlooked. In addition, damage tends to be patchy and the plant stand is thinned or weakened rather than destroyed altogether. Probably, however, the main reason why damage passes unheeded is that there is usually no uninfested area with which to compare the (normal) infested one. In a recent assessment of losses caused by pests to establishing swards spring sowings were not frequently affected – less than one-fifth of the sites at which studies were made. However, two-thirds of autumn sowings were significantly damaged by pests (10). Total failures as a result of pest attack which require a further attempt at re-sowing probably account for 1–2% of newly-sown swards.

Several factors accentuate damage. For example, time of sowing – there is much greater risk of damage by frit fly to autumn-sown grass. Swards sown from mid to late August seem especially at risk – over 40% of tillers may be destroyed. Re-seeds where grass follows grass rather than an arable crop seem more prone to damage. Large numbers of some pests build up in old swards and this reservoir of infestation then attacks the new sowing. Use of direct-drilling and probably other minimum tillage methods greatly increases the risks. Desiccating an old pasture with herbicide dries up the food source for frit fly larvae, which then migrate on to the newly-sown seedlings. Slugs also find the slots made by some direct-drilling machines an ideal environment.

The likelihood of pest damage to a newly sown pasture can be minimized. Ploughing and other cultivations associated with conventional seed-bed preparation damages many pest larvae and exposes them to desiccation and predation. Creating a fine tilth and ensuring good consolidation by heavy rolling removes crevices in which slugs may lurk and physically restricts movement by slugs, leatherjackets and wireworms.

Certain pesticides (see Chapter 3) are approved for the control of establishment pests. In view of the risks involved especially to autumn sowings their use should be considered. The application of pesticides to newly seeded areas (in autumn) is likely to produce considerable benefits during the early phases of establishment which may also be reflected in further long-term yield differences.

Seed treatments and/or a range of granular pesticide treatments are likely to become available for the control of establishment pests in future.

Diseases

Grassland is subject to some 400 diseases on a world basis, of which over 150 have been recorded in the UK (5). Some are seed-borne (21); the majority are caused by fungi (23), the remainder by viruses and mycoplasma, with a few bacterial diseases. Most of the viruses and mycoplasma are transmitted by vectors of one kind or another. Some of these vectors are themselves fungi but the vast majority are animal in nature: aphids, beetles, plant- or leaf-hoppers, or eriophyid mites (9). Aphids in particular thus have a dual role, as pests in their own right and as transmission agents of important grass and clover viruses.

Less than 40% of seeds of sown grasses actually contribute plants to the sward, even in the early establishment phase (19). Much of this failure is due to diseases of the pre- and post-emergence damping-off type, and thus represents an avoidable loss. Subsequently, other pathogens may become prevalent, varying in intensity with the maturity of the sward, edaphic, climatic and management factors. All parts of the plants may become infected, the leaves most commonly so. On most well established plants this may lead simply to a temporary loss of yield at a particular grazing or conservation cut, although the effects of disease may be rather more persistent. For example, a heavy infection of crown rust on ryegrass in early autumn reduces regrowth in spring, even in the absence of any apparent infection, by as much as 20%. Equally, infected plants may lose competitive ability, and hence be unable to maintain an effective contribution to the sward. Viruses, which unlike fungi are generally systemic throughout the plant, often tend to modify growth habit, altering growth rhythm and markedly affecting response to management as well as influencing competitive ability (7).

Precise losses in grassland due to diseases are poorly documented and difficult to estimate, largely because, with so complex a crop, having an indirect output through animal production, the influence of disease has to be considered against the many other factors and types of management influencing yield potential. Few grassland areas have been exploited to peak capacity, and the main emphasis has been on increasing output by intensifying management systems. As production levels come closer to the limit of crop and cultivar potential, so disease is moving nearer to becoming a dominant factor limiting a further increase in production. Estimates in North America indicate losses of 6% annually from individual diseases and 12% over all diseases combined (4). In Britain, whenever infection is at all heavy, losses can be of the order of 35% for crown rust, 20% for both powdery mildew and rhynchosporium leaf blotch, and 15% for ryegrass mosaic virus. However, measurable losses in total yield represent only part of the story. Cattle tend to reject badly rusted herbage totally, reducing effective intake. Some common fungal pathogens markedly affect nutritive quality by reducing soluble carbohydrate, carotene or crude protein content, and thus increasing fibre. In addition to the digestive upsets which may be caused to the grazing animal by severely diseased herbage, some legume pathogens may also cause a marked rise in the concentration of oestrogenic compounds, adversely affecting reproduction.

Grassland diseases rarely reach the epidemic levels commonly noted in cereals infected with rusts or mildew. New grass is commonly sown as a mixture of species, together with a legume companion, while permanent pasture almost invariably consists of a mixture which has approached equilibrium. There is therefore considerable buffering against the emergence of a single pathogen epidemic; if one species within a mixture becomes badly infected another, immune or at least less susceptible, tends to take its place (6). Modern trends towards monoculture in short term leys, although enhancing overall as well as specific yield potential, reduce this buffering effect and so exacerbate disease risk. This was noticeable at the time when Italian ryegrass was sometimes grown in a continuous cropping sequence with heavy nitrogen dressings replacing the traditional legume companion. Take-all resulted on an unprecedented scale together with an increased incidence of leaf diseases. Even with mixtures, particularly simple ones, sward balance can still be affected: infection in cocksfoot/white clover swards with the lethal cocksfoot mottle virus has led to virtual total dominance of the white clover (7). Italian ryegrass has also been observed to disappear from young leys owing to its greater susceptibility than perennial ryegrass to the root pathogen, *Ligniera junci* (20). The cocksfoot constituent of mixed swards has increased when the ryegrass component correspondingly diminished through infection with *Drechslera siccans* severe enough to induce a foot rot as well as the usual leaf spot symptoms. In all of these cases the resultant changes in sward composition may well have led to variation in seasonal yield distribution as well as a reduction in total yield. In extreme cases, affected plants have been overtaken by weeds. Such observations are well documented in the eastern half of Britain.

3

Protection against epidemics has arisen in the past by the development of cultivars having a broad genetic base (grasses, unlike cereals, consist of populations of outbreeding individuals) from indigenous sources in which the plants have become largely adapted to the pathogens to which they have been exposed. The extent of disease present in a plant introduction nursery illustrates the extent to which natural selection has operated in keeping indigenous grasses relatively disease free. However, in achieving greater uniformity or a more precisely defined function modern cultivars tend to have a more limited genetic base and to be less buffered against disease. The situation is further exacerbated by the introduction of unadapted, non-indigenous material, both in the raw state by the plant breeder in the legitimate search for new gene sources and also in the form of cultivars imported from abroad. An example of the former is the use of ryegrass derived from the Po valley region of Italy, which proved so susceptible to the normally minor pathogen brown rust, that considerable selection was necessary to bring it to an acceptable level of resistance, and of the latter, the Swiss bred Italian ryegrass cultivar, Lior, which was ultimately withdrawn by the breeder because of its extreme susceptibility to crown rust and drechslera leaf spot (22) and also to mildew and rhynchosporium leaf blotch (11).

Although plant pathologists tend to study the effects of each disease in isolation, it is not uncommon in the natural situation for crops to be infected with more than one pathogen at the same time. Where these interact, either by each cancelling the effects of the other or by mutual enhancement, the problems of assessment and control may become even more complex. For example ryegrass mosaic virus, which reduces soluble carbohydrate, can suppress the development on ryegrass of the high-carbohydrate-requiring crown rust (18), whereas barley yellow dwarf virus, which causes carbohydrate accumulation, has something of the opposite effect. Ryegrass plants infected with barley yellow dwarf virus and with a mild isolate of ryegrass mosaic virus exhibited the dwarfing due to the former but not the enhanced tillering, which was in fact reduced by the action of the latter virus (8).

Grassland diseases may be controlled through chemical agents, management practices, and the development and use of resistant cultivars. An integrated approach, involving all three strategies as appropriate, probably offers the best solution.

Chemical control is most likely to be effective during establishment, particularly when applied as seed dressings (see p. 29), and may also be desirable to check the spread of important viruses when a common pest is also the vector (see p. 69). Otherwise, chemical control is probably rarely justified on an economic basis except for seed production where the value of the cash crop is high enough to merit it. In consequence, most research on the use of fungicides has been based on small scale glasshouse or field plot trials and cannot at present be readily translated to the field situation.

The scope for disease control through management is also limited, as overall management must include the need to maximize yield, often irrespective of disease. In practice, good husbandry is often the best insurance. Allowing the crop to reach over-maturity, particularly at the end of the season, will increase disease levels, both directly and by allowing the build-up of spores and other forms of inoculum for subsequent re-infection. Where there is a high disease level it may help to cut or graze earlier. Identification and a thorough understanding of a disease may well enable specific cutting or grazing regimes to be applied to minimize disease loss. Good drainage is important in preventing many of the root rots and the build-up of largely saprophytic fungi which can cause considerable spoilage under wet conditions.

Losses can sometimes be offset by application of nitrogen, though here there are distinct complications. For example, crown rust is said to be reduced by nitrogen application, whereas drechslera leaf-spots tend to become more severe (17), while swards infected with ryegrass mosaic virus notably fail to respond to nitrogen treatment (1). Manipulation of the sowing date can sometimes avert disease. For example, because they avoid the late spring/early summer peak of mite vectors, autumn-sown ryegrasses are less prone to reygrass mosaic virus infection than are those sown in spring.

The best long-term prospect lies in developing disease resistant cultivars. Considerable progress in this direction has already been made by plant breeders, with a number already available and more to follow. The recent inclusion by the National Institute of Agricultural Botany (NIAB) of disease ratings for grass cultivars on their recommended list (2) to accompany the information on herbage legume diseases which has been available for a number of years from this source (3), should help farmers to choose those which will satisfy their particular needs.

References

1. A'Brook, J.; Heard, A. J. (1975). The effect of ryegrass mosaic virus on the yield of perennial ryegrass swards. *Annals of Applied Biology,* **80,** 163–168.
2. Anon (1983*a*). Recommended varieties of grasses 1983/84. Cambridge: National Institute of Agricultural Botany, Farmers' Leaflet No. 16, 23 pp.
3. Anon (1983*b*). Recommended varieties of herbage legumes 1983/84. Cambridge: National Institute of Agricultural Botany, Farmers' Leaflet No. 4, 15 pp.
4. Berkenkamp, B. (1974). Losses from foliage diseases of forage crops in central and northern Alberta, 1973. *Canadian Plant Disease Survey,* **54,** 111–115.
5. Carr, A. J. H. (1971). Herbage legume diseases; Virus diseases of forage legumes; Grass diseases. In *Diseases of crop plants,* pp. 254–307, Western, J. H. (Ed.), London: MacMillan, 404 pp.
6. Carr, A. J. H. (1979). Causes of sward change – diseases. In *Changes in sward composition and productivity,* pp. 161–166, Occasional Symposium No. 10, British Grassland Society, Charles, A. H.; Haggar, R. J. (Eds.), 253 pp.
7. Catherall, P. L. (1966). The significance of virus diseases for the productivity of grassland. *Journal of the British Grassland Society,* **21,** 116–122.
8. Catherall, P. L. (1979). Virus diseases of cereals and grasses and their control through plant breeding. *Welsh Plant Breeding Station Annual Report for 1978,* 205–226.
9. Catherall, P. L. (1981). Virus diseases of grasses. *Ministry of Agriculture, Fisheries and Food Leaflet 595,* 8 pp.
10. Clements, R. O.; French, N.; Guile, C. T.; Golightly, W. H.; Lewis Serfiah; Savage, M. J. (1982). The effect of pesticides on the establishment of grass swards in England and Wales. *Annals of Applied Biology,* **101,** 305–313.
11. Davies, H.; Williams, A. E.; Morgan, W. A. (1970). The effect of mildew and leaf blotch on both yield and quality of *Lolium multiflorum* (cv. Lior). *Plant Pathology,* **19,** 135–137.
12. Dibb, C.; Haggar, R. J. (1979). Evidence of effect of sward change on yield. *ADAS Quarterly Review,* **32,** 1–14.
13. Doyle, C. J. (1982). Economic evaluation of weed control in grassland. *Proceedings 1982 British Crop Protection Conference – Weeds,* 419–427.
14. Forbes, J. C.; Kilgour, D. W.; Carnegie, H. M. (1980). Some causes of poor control in *Senecio jacobaea* by herbicides. *Proceedings 1980 British Crop Protection Conference – Weeds,* **2,** 461–468.
15. Haggar, R. J.; Elliott, J. G. (1978). The effect of dalapon and stocking rate on the species composition and animal productivity of a sown sward. *Journal of the British Grassland Society,* **33,** 23–33.
16. Kirkham, F. W.; Haggar, R. J.; Elliott, J. G. (1982). Controlling weeds during grass establishment. *Weed Research Organization 9th Annual Report,* 55–61.
17. Lam, A.; Lewis, G. C. (1982). Some effects of nitrogen and potassium fertilizer application on *Drechslera* spp. and *Puccinia coronata* attack on perennial ryegrass (*Lolium perenne*) foliage. *Plant Pathology,* **31,** 123–131.

18. Latch, G. C. M.; Potter, L. R. (1977). Interaction between crown rust (*Puccinia coronata*) and two viruses of ryegrass. *Annals of Applied Biology,* **87,** 139–145.
19. Michail, S. H.; Carr, A. J. H. (1966a). Effect of seed treatment on establishment of grass seedlings. *Plant Pathology,* **15,** 60-64.
20. Michail, S. H.; Carr, A. J. H. (1966b). Italian ryegrass, a new host for *Ligniera junci. Transactions of the British Mycological Society,* **49,** 411–418.
21. Noble, M.; Richardson, M. J. (1968). An annotated list of seed-borne diseases. *Commonwealth Mycological Institute, Kew, Surrey, Phytopathological Papers No. 8,* 191 pp.
22. Nüesch, B. (1964). "Renova" und "Lior", zwei neue Futter Pflanzensorten. *Mitteilungen für die Schweizerische Landwirtschaft, Zurich,* **12,** 1–12.
23. O'Rourke, C. J. (1976). *Diseases of grasses and forage legumes in Ireland.* Carlow, Eire: The Agricultural Institute, Oak Park Research Centre, 115 pp.
24. Peel, S.; Hopkins, A. (1980). The incidence of weeds in grassland. *Proceedings 1980 British Crop Protection Conference – Weeds,* **3,** 877–890.
25. Spedding, C. R. W. (1966). Weeds and animal productivity. *Proceedings 8th British Weed Control Conference,* **3,** 854–860.

Chapter 2   INTEGRATED CONTROL OF WEEDS, INSECTS AND PATHOGENS

The term "integrated control" came to the fore in the late 1940s when entomologists were concerned with the build up of resistance in insect populations being sprayed with insecticides. It became obvious that a more rational approach to the use of insecticides was needed and new control systems were developed, based on sound ecological principles, which proved to be less expensive and environmentally more acceptable. Thirty years later, the widespread use of prophylactic spraying with herbicides in arable crops is forcing weed scientists to adopt a similar course of action and "integrated pest (weed) management" has become a fashionable term, although often misused, misunderstood and even maligned.

For the purpose of this book, integrated control is defined as a management system that, based on a knowledge of the ecology and population dynamics of the particular organism, uses all appropriate techniques (including cultural, chemical and biological) in as compatible a manner as possible – and with due consideration to environmental quality – to maintain populations of these organisms at levels below those causing economic damage in the current and future years. This approximates to the FAO definition of integrated pest management as outlined by Geissbuhler (5).

Grassland farmers have for years been using cultural methods ("good management") to avoid or suppress weeds and insect pests, realizing that the best form of control is a dense vigorously growing sward. Thus, they have long since been aware of the importance of cultural techniques like sound crop rotation; sensible choice of well-adapted species (including multiple pest-resistant crop varieties and herbicide-tolerant varieties); effective cultivation and seedbed preparation; optimum plant populations; adequate fertilizer application; appropriate and controlled grazing methods. Even when selective herbicides became available they were not taken up as readily as in arable farming – for a variety of reasons including the high cost of the chemical, lack of appropriate spraying equipment, difficulty in identifying weed infestations (swards often look deceptively dense 6 months after sowing), lack of evidence on the damage caused by weeds, plus suspicions about the efficacy of certain herbicides.

In recent years, however, the improvement in the selectivity of herbicides has made it possible to treat grass more like an arable crop, such as a cereal, where weed control is the rule and emphasis is given to optimizing the performance of the crop. (For instance, there is now more need to use grass seed rates to optimize sward productivity, rather than to use high seed rates to control weeds during establishment.)

Once weed-free swards have been established, they are still likely to come under stress at some stage or other, not least from the environment, leading to ingress and attack by weeds, pests and diseases. Often this will mean recourse to mechanical control methods, e.g. mowing or hoeing, although these are often ineffective and only relevant to isolated infestations.

Where these cultural and mechanical methods are only partly successful chemical control methods may be needed, including selective herbicides, insecticides and fungicides. Additionally, some biological control methods are becoming available including the use of pest-attacking predators, parasites and pathogens. (Biological control is defined as the regulation of the population of a living organism – insect, weed, etc. – by another living organism, known as its natural enemy – predator, parasite or pathogen.)

As options and alternatives (see Table 2.1) continue to increase, and because they all impinge one with the other, so the various separate disciplines and interests need to be brought together into a systems approach to habitat management, the overall objective being to maximize the pressure that the growth of grass (and/or clover) can exert on the competing organisms.

TABLE 2.1 Summary of non-chemical methods of controlling weeds, pests and diseases

Organism	Cultural	Mechanical	Biological

Weeds	Crop rotation
	Rectifying site
	  deficiencies
	Choice of species
	Time of sowing
	Use of companion crop
	Seed rate
		Ploughing/hoeing
		Harrowing/discing
		Mowing
		Grazing
		Burning
			Weed-eating insects,
			  mites and pathogens

Insects	Crop rotation
	Time of sowing
	Time of harvesting
	Trash/stubble
	  management
	Planting border plants
	Destruction of
	  alternate hosts
	Sanitation
	Soil, water, fertilizer
	  use
		Ploughing
		Rolling
		Grazing
			Insect predators,
			  parasites, microbial
			  and viral preparations
			Breeding resistance

Diseases (fungal, bacterial and viral)	Crop rotation Sanitation Seed treatment Trash management	Defoliation	Microbial pathogens

Advances in Cultural, Mechanical, Chemical and Biological Control

Weeds

Weeds, particularly broad-leaved species, by virtue of their size, shape and colour, are more self-evident than insects and diseases and so weed control technology is more developed in grassland than insect- or disease-control technologies.

Cultural methods of curtailing weed ingress in newly sown grassland, by the establishment of a dense vigorously growing sward, are outlined in Chapter 3.

In established swards, botanical degeneration frequently reflects lack of correct management. Pioneer species like rushes and thistles can be prevented from establishing if a dense ground cover is maintained by rotational (rather than continuous) grazing, mowing, fertilizing and periodic reseeding. It needs to be noted, however, that fertilizing with nitrogen does not always work satisfactorily since it can lead to an increase in fertility demanding weeds like common chickweed (*Stellaria media*), annual meadow-grass (*Poa annua*) and docks (*Rumex* spp.).

If cultural methods fail to prevent weed invasion, alternative mechanical methods can be considered, including strategic mowing and even hoeing. However, these are impracticable on certain terrains and where weed densities are high. In this case, chemical control is needed.

Herbicides like MCPA and 2,4-D have been available for killing harmful broad-leaved weeds in grassland since the 1950s. However, adoption of their use by farmers was restricted, often because of the harmful side effects of these herbicides on white clover. Later, clover-safe herbicides were developed, e.g. MCPB; 2,4-DB; benazolin; asulam, although their greater cost, and the availability of cheap nitrogen, limited their widespread use in grassland. Perhaps of even greater importance, however, was the lack of compelling evidence on the need to control these weeds, especially non-poisonous species, in permanent swards.

In recent years, as costs continue to rise and with the continued trend towards increasing outputs from grassland, more interest has centred on problems caused by specific weeds, especially those associated with newly sown crops e.g. common chickweed and annual meadow-grass, and with intensively managed leys, e.g. docks. Also, there has been a trend, dictated by cost considerations and the availability of new chemicals, for herbicide-based methods to replace the more conventional cultural and mechanical methods of establishing grass. This trend is likely to continue as costs rise.

Nevertheless, a clear distinction needs to be made between the empirical use of herbicides as a substitute for mechanical and cultural weed control and the systematic use of herbicides, in conjunction with appropriate cultural and husbandry practices, based on a scientific understanding of the long-term responses of the weeds to these particular inputs, the latter approach being referred to as "weed management". A further refinement, implicit in the term "integrated weed management" (4) is that the control of weeds should, if feasible, be integrated with requirements for protecting crops additionally from insect pests, pathogens, nematodes and other injurious organisms.

All weeds have natural enemies, principally insects, mites and pathogens. A recent review (10) shows that biological control by the introduction of exotic natural enemies for the control of introduced weeds is often practical and can be extremely effective against single weed species that infest large areas of grassland, while the manipulation of native natural enemies, especially pathogens, for the control of indigenous weeds is a rapidly expanding field.

Thus, most North American thistle weed problems are of European or Eurasian origin and there are extensive research programmes under way for the biological control of *Carduus* spp. and *Cirsium* spp. by the introduction of host-specific insects from Europe (9). The introduction of the seedhead feeding weevil, *Rhinocyllus conicus* has led to substantial reduction of the incidence of *Carduus nutans* where introduced in the US and Canada (6). Although one natural enemy, the European beetle *Altica carduorum*, whose larvae feed on the roots of *Cirsium arvense*, was released in England in 1969, this was more of a scientific experiment than a real attempt at biological control, as the beetle appears to be restricted to areas where temperatures do not fall below 20°C for several months of the year. So, as yet, the method of "classical" biological control by the introduction of exotic natural enemies has still to be taken up seriously in the UK. However, investigations have been initiated to locate suitably host-specific natural enemies for the biological control of bracken (*Pteridium aquilinum*). Bracken is distributed world wide in suitable habitats and its associated natural enemies differ from area to area. The types of insects associated with bracken in Europe and North America differ markedly from those found in the highlands of Papua New Guinea and South Africa. J. H. Lawton (personal communication) has found a stem boring moth, *Parthenodes angularis*, which causes substantial damage to bracken in South Africa, and in co-operation with the Commonwealth Institute of Biological Control hopes to demonstrate that it is sufficiently host-specific to be considered for introduction into the UK.

The release of mass reared indigenous weed feeding insects is unlikely to prove a useful technique in pasture management, but the application of mass-produced, indigenous, host specific fungi has already been shown to be effective. A 1981 catalogue (11) lists 90 such research projects, mostly from the US, of which five were operational, 26 were being field tested and 32 were under preliminary study. In Scotland, work on pathogen control of bracken is in progress (M. N. Burge personal communication).

At Oxford, preliminary studies in biological control of weeds by fungi are showing that low doses of 2,4-D increase rust disease in creeping thistle (M. P. Greaves, personal communication).

Insects

Many insect species live in grassland and some are extremely numerous. Occasionally, populations may "explode" causing a noticeable change in sward composition (1). Young swards are especially vulnerable to insect attack. For example, newly sown Italian ryegrass is particularly susceptible to attack by frit fly (Oscinella frit). More often, however, less apparent but more widespread and continuous damage is caused by invertebrate fauna at lower, endemic population levels.

Cultural measures to prevent or suppress specific insect populations (such as crop rotation; sanitation; destruction of alternative host plants, trash or stubble; fertilizer practice etc.) are often feasible. Of particular importance is timeliness of sowing, which can have a profound effect on the likelihood of pest attack; the reasons for the increased risk of pest attack under many circumstances are understood and consequently control measures may be devised. As another example, insect-induced deterioration of ryegrass swards can be avoided by sowing species like timothy, cocksfoot and meadow fescue which are virtually immune to frit fly attack.

Killing insects mechanically is rarely feasible, except when soil is disturbed by ploughing. However, certain larvae, e.g. cranefly (Tipula paludosa) and slugs can be killed by rolling or animal treading.

Insecticides provide the main method of controlling damaging insects. For establishing grassland the use of insecticide seed dressings has shown considerable promise. Various insecticides can also be used on established grassland, notably the non-persistent chlorpyrifos.

Biological control is also a possibility and has been used very successfully, for example in the US where there are two outstanding results against introduced pests; the Rhodes grass scale (Antonina graminis) was suppressed by a parasitic wasp (Neodusmetia sangwani) imported from India (3) and the Japanese beetle (Popillia japonica) is controlled by application of spores of milky disease (Bacillus popilliae) (2).

Diseases

Fungal pathogens are endemic in most soils and can cause significant losses to grass seedlings. In established grassland, certain viruses can reduce sward productivity and response to fertilizers.

The present use of cultural practices to suppress or avoid pathogen attacks in grassland (including sanitation; disease-resistant varieties; seed treatment, timeliness of sowing; trash management) is rather limited. Chemical control, too, is rarely feasible, whilst biological control methods (by the manipulation of microbial pathogens or antagonists against fungi, bacteria and nematodes) have only just begun to be investigated (2). This leaves frequent defoliation as the main method of preventing foliar pathogens reaching epidemic proportions.

Need for Compatibility of Control Techniques

In any integrated crop protection management system involving more than one form of control, the effect of the various control components on the efficiency of each other needs to be noted (see Fig. 2.1). For instance, using fungicide-treated seed might affect white clover nodulation, leading to reduced crop vigour and, hence, poorer weed control.

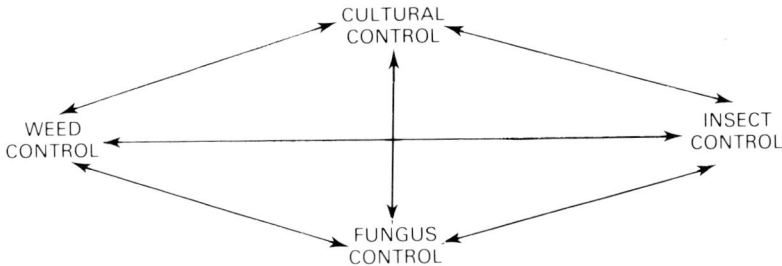

Fig. 2.1   Interactions of cultural and chemical control agencies.

When crop protection is widened to include integrated pest management (see Fig. 2.2) there is even more risk of one control component reducing the efficiency of another. For instance, spraying thistles with 2,4-D (chemical control) can have a damaging impact on populations of beneficial weed-eating insects (biological control). In this case, timing of 2,4-D application needs to be adjusted to have a minimal impact on the insect populations (12).

As yet biological control involving the widespread release of weed-eating insects has not become a practical reality in the UK, although it is known that many of the major weeds of grassland are attacked by phytophagous insects, e.g. bracken (*Pteridium aquilinum*) by *Phyllopertha horticola*, ragwort (*Senecio jacobaea*) by the moth *Tyria jacobaeae*, docks (*Rumex* spp.) by the beetle *Gastrophysa viridula*. These native insects help keep these native weeds under control – which is why some of them have been introduced to other parts of the world. With this knowledge, therefore, it would be prudent, by modifying spraying times, to avoid damaging these predators. For the future, the compatibility of chemical and biological agents will need to be tested during the early stages of an integrated control programme (Fig. 2.3).

Unfortunately, little is known about the compatibility of cultural and biological control agents. Disturbance of the sward (by ploughing, spraying or grazing) can interrupt the developmental cycle of the biological agent and so impair control. This indicates that the use of weed-killing insects in newly sown grassland may be more difficult to develop than for undisturbed permanent grassland.

Cultural control techniques are not always compatible with chemical control approaches. For example, using close grazing and the oversowing of legumes in New Zealand to control bracken (*Pteridium esculentum*) restricts herbicide choice, whilst close grazing of ragwort may make this species more difficult to kill chemically (7).

11

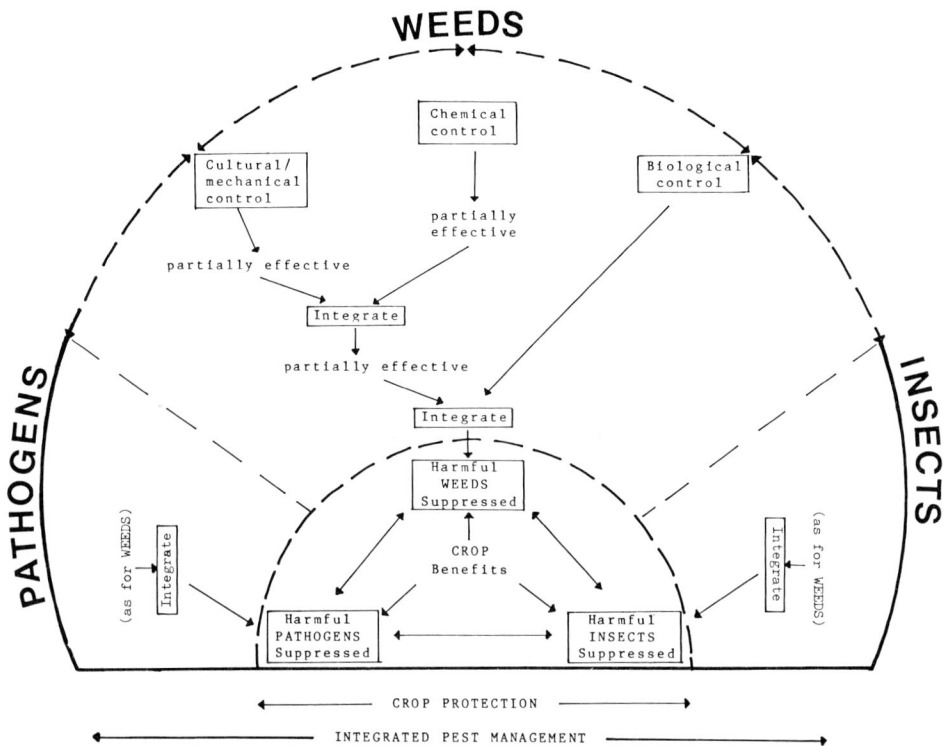

Fig. 2.2   Interactions in an integrated system of weed, pest and disease control.

Interactions between Weeds and Other Pests

Any type of control measure, be it cultural, mechanical, chemical or biological, which causes changes in vegetation can indirectly alter the habitat and hence the populations of insects, pathogens, nematodes within that vegetation (see Fig. 2.4); these changes in populations may be beneficial, detrimental or of no significance to crop performance. For instance, because weeds provide a habitat for both beneficial and pest insects (or pathogens) weed control measures will indirectly alter the survival of these dependent organisms, with good or bad effects. Norris (5) gives examples of beneficial weed/insect associations and points out that disturbance of these associations can sometimes damage food sources of wild game, e.g. partridges. On the other hand, however, weeds can increase insect pest problems – not least by acting as alternative hosts – and encourage the spread, or carry over, of diseases, e.g. transmission of ergot from weedy grasses back to cereals.

Interactions can also occur directly by chemicals killing non-target organisms such as insects, fungi and nematodes. Alternatively, they may modify the physiology of the host plant such that insect or pathogen growth is changed. For instance, spraying grass with certain selective herbicides often increases the risk of disease attack (R. J. Hall, personal communication).

In practice, most weeds occur as complexes of several species. Hence, killing or suppressing the target weed species may not necessarily lead to improved crop growth,

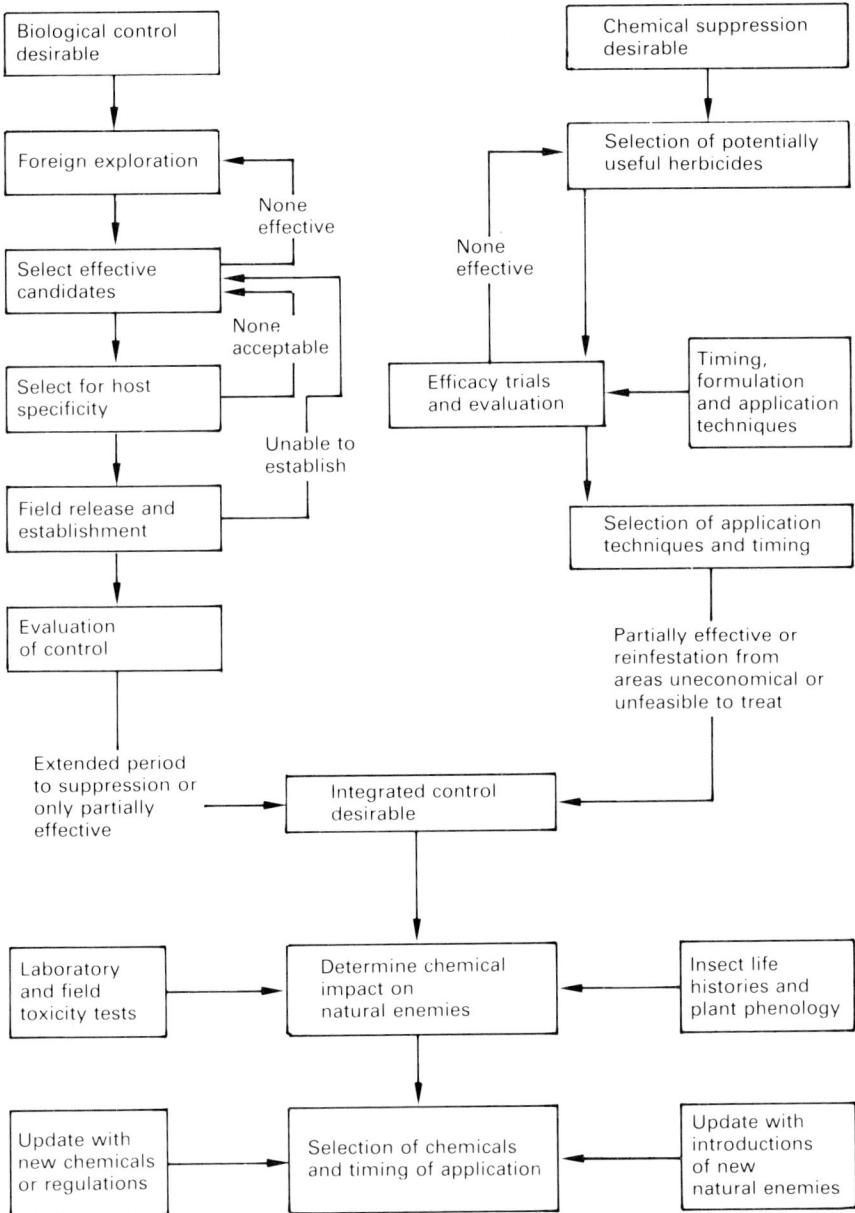

Fig. 2.3 A flow diagram of biological and chemical control techniques and their integration for developing pest management systems for thistles. From J. T. Trumble and L. T. Kok, Integrated pest management techniques in thistle suppression in pastures of North America. Reproduced, with permission, from *Weed Research*, Volume 22. © Blackwells Scientific Publications.

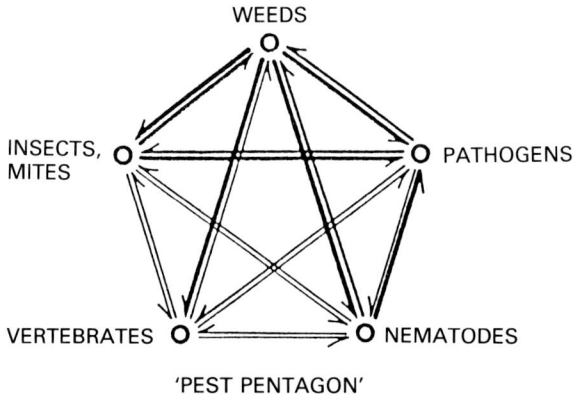

'PEST PENTAGON'

Fig. 2.4 A diagrammatic presentation indicating the potentials for interactions between different types of pest organisms. From R. F. Norris, Interaction between weeds and other pests in agro-ecosystems. Reproduced, with permission, from *Biometeorology in Integrated Pest Management,* edited by J. L. Hatfield and I. J. Thomason. © 1982 Academic Press, Inc.

especially if the remaining weed species become more vigorous. For example, the benefits of killing chickweed in newly sown perennial ryegrass with mecoprop can be undermined by increased vigour of surviving annual meadow-grass released from competition by the chickweed.

Additive Effects of Control Techniques

When control techniques are fully compatible, and do not lead to harmful interactions, improved results can be expected, compared with results from the control techniques used separately. For instance, killing weeds in the seedbed by cultural means can, by exposing mineral soil, increase the efficiency of soil-acting herbicides. Also, significant additive effects of ethofumesate (chemical control) and seed rate (cultural control) have been reported in newly sown swards (R. O. Clements, personal communication). Another example of cultural and herbicidal methods working for mutual advantage is the use of burning/grazing pressure and spot applications of herbicides for brush control in New Zealand 'hill' country (7). Similarly, the palatability of nettles, and even thistles, is increased by spraying, whilst the combined use of spraying and grazing is of special use in controlling seedlings of these species. Yet a further example is the combined effect on thistles of phytophagous insects and synchronized spraying of 2,4-D being greater than that produced by either agency used separately (12).

Chemical control techniques involving more than one chemical can also lead to additive effects. For example, in one experiment involving nine pesticide and fungicide treatments applied to permanent pasture (R. O. Clements, personal communication) none of the treatments enhanced yield significantly when used alone. However, large responses in herbage yield occurred with the combined pesticide and fungicide treatment. In another investigation, a three-way combination of herbicide (ethofumesate) + insecticide (phorate) + fungicide (benomyl) has been reported (R. D. Sheldrick, personal communication) as giving improved yield of sown grass compared with each of the component chemicals used on its own.

14

Conclusions

The integrated use of crop protection chemicals makes farmers less dependent on cultural methods of weed, pest and disease control, thus making it possible to optimize crop performance. In future, biological control agents could well increase in importance, especially when integrated fully with cultural and chemical controls. To be effective and acceptable, however, such integrated pest management systems must be economic, technically sound and practically reliable. More knowledge will be needed to improve farmers' decisions in choosing methods for use in controlling complexes of weeds, pests and diseases (noting that busy farmers are often more concerned with treating symptoms rather than preventing causes). In particular, more information is needed on economic threshold densities of weeds, predicting/monitoring insect population, pathogen dynamics etc.

Meanwhile, cultural methods, including the use of vigorous grasses and clovers (bred for resistance to weeds, pests and disease attack), will remain the cornerstone for controlling these contaminants. Chemical methods of control, however, are likely to expand and increase in sophistication, although they could be undermined by a build up of varietal resistance, and by increased pressure from the anti-pesticide lobby.

References

1. Clements, R. O.; Henderson, I. F. (1979). Insects as a cause of botanical change in swards. In *Changes in sward composition and productivity,* pp. 157–160. British Grassland Society Occasional Symposium 10. Charles, A. H.; Haggar, R. J. (Eds.).
2. Deacon, J. W. (1983). *Microbial Control of Plant Pests and Diseases.* Wokingham: Van Nostrand Reinhold, 88 pp.
3. Dean H. A.; Schuster, M. F.; Boling, J. C.; Riherd, P. T. (1979). Complete biological control of *Antonina graminis* in Texas with *Neotusmetia sangwani. Bulletin of the Entomological Society of America,* **25,** 262–267.
4. Fryer, J. D. (1981). Weed management: fact or fable? *Philosophical Transactions of the Royal Society of London, Series B,* **295,** 185–197.
5. Geissbuhler, H. (1981). The agrochemical industry's approach to integrated pest control. *Philosophical Transactions of the Royal Society of London, Series B,* **295,** 111–123.
6. Harris, P. (1984). *Carduus nutans* L., Nodding Thistle and *C. acanthoides* L., Plumeless Thistle (Compositae). In *Biological Control Programmes against Insects and Weeds in Canada 1969–1980,* pp. 115–126. Kelleher, J. S.; Hulme, M. A. (Eds.) Farnham Royal: Commonwealth Agricultural Bureaux.
7. Matthews, L. J. (1977). Integrated control of weeds in pastures. In *Integrated Control of Weeds,* pp. 88–119. Fryer, J. D.; Matsunaka, S. (Eds.) Tokyo: University Press.
8. Norris, R. F. (1982). Interactions between weeds and other pests in the agro-ecosystem. In *Biometeorology in Integrated Pest Management,* pp. 343-406. Hatfield, J. L.; Thomason, I. J. (Eds.) New York: Academic Press.
9. Schroeder, D. (1980). The Biological control of thistles. *Biocontrol News and Information,* **1,** 9–26.
10. Schroeder, D. (1983). Biological control of weeds. In *Recent Advances in Weed Research,* pp. 41–78. Fletcher, W. W. (Ed.) Farnham Royal: Commonwealth Agricultural Bureaux.
11. Templeton, G. E. (1981). Status of weed control with plant pathogens. In *Biological Control of Weeds with Plant Pathogens,* pp. 29–44. Charudattan, R.; Walker, H. L. (Eds.) New York: John Wiley & Sons.
12. Trumble, J. T.; Kok, L. T. (1982). Integrated pest management techniques in thistle suppression in pastures of North America. *Weed Research,* **22,** 345–359.

Chapter 3 THE ESTABLISHMENT PHASE

The period during which the sown grass and clover species are germinating and developing into seedlings is of vital importance, because what happens during this establishment phase determines the composition and nature of the sward that is to be formed, and consequently its potential yield. If the crop which is sown does not become well established, it will be replaced by a 'crop' of weeds, forming a sward which may appear to be of reasonable quality, but which will not have the productive capacity intended. If a sward sown in September is infested by grass weeds its yield can be reduced by 18–45% by the following June. It is therefore as necessary to aim at obtaining good establishment of a grass (or grass/clover) crop as it is for any other crop.

Successful establishment depends first on good germination – a high percentage of seeds germinating rapidly and evenly to produce vigorous seedlings. This will only occur if seed of good quality in terms of germination capacity, vigour and uniformity is sown under soil conditions conducive to good germination. It is easier, by using certified seed with a high 1000-seed weight, to ensure seed quality than ideal soil conditions, but a great deal can be done to improve conditions even if they cannot be made ideal.

The site needs to be adequately drained, the soil pH near 6.0 and the supply of mineral nutrients adequate. To ensure this, lime, nitrogen, phosphate and potassium may all need to be applied to the seed bed. Nitrogen is more effective if applied soon after emergence, but it is usually more convenient to apply it to the seed bed; phosphates are immobile in the soil and are best placed below the seed; potassium is more mobile and can increase vigour of establishing seedlings. The amounts necessary depend on their status in the soil already, but unless they are known to be deficient or very high, 30–50 kg ha^{-1} of N, 100–120 of P and 40–80 of K should be adequate.

The physical state of the soil is also important. The necessity for a good tilth, however, is more generally agreed than is its definition. A good crumb structure, ensuring adequate aeration, drainage and evenness in depth of sowing, is important; a high content of organic matter is better avoided unless it has been well leached, since it may contain toxins produced by the decay of dead plant material, as well as harbouring pests and diseases.

The amount of seed sown affects establishment less than might be expected. With ryegrass, an ideal to aim for is about 250 plants, with 6000 tillers, per square metre, in the spring following sowing. Under ideal conditions, with 80% emergence, 5–10 kg ha^{-1} of seed can give 200–250 plants per m^2, so a sowing rate of 10 kg ha^{-1} can give good establishment. Farmers often sow at rates as high as 30 kg ha^{-1} in order to allow a safety margin, an argument in favour of this practice being that the cost of the extra seed is small compared with that of reseeding if a crop fails. However, the advantage is doubtful, since good germination can give an excessive density of seedlings, which will cause diseases to flourish. It has also been claimed, however, that on average, sowing under field conditions and using a normal seed rate, some 75% of viable seeds sown fail to produce plants which survive the first two months and this can increase to 90% after the first winter. Variation of seed rate between very wide limits has little effect on dry-matter production, since the total number of tillers is much the same, whether the stand consists of densely packed plants with few tillers each or more widely spaced ones with many. On the whole, it would seem advisable to use as low a seed rate as possible (especially when clovers are included in the seed mixture), since plants developing at lower densities will suffer less from competition, develop better root systems and more tillers; those that survive from a very dense initial stand may not be the most desirable type of plant.

The seed needs to be covered, the optimum depth for perennial ryegrass being 2–3 cm; intimate contact with the soil needs to be ensured by rolling, which improves the moisture relations and thus evenness of germination.

The weeds, pests and diseases already present will be on the whole those of the previous crop, although seeds of many weed species can remain dormant in the soil for many years and be stimulated into germination by cultivation. Consideration is given in the remainder of this chapter to those organisms most likely to give trouble during the establishment period and the means of dealing with them. Prevention is better than cure, however, and the best method of preventing these troubles is to achieve good, rapid establishment of a strong, healthy sward, subsequently managing it to keep it so.

Controlling Weeds During Establishment

Weed occurrence and impact

Seedbed preparation and drilling stimulate weed seeds to germinate. The actual species which occur vary depending on previous cropping (13); young leys in an arable rotation are commonly infested with common chickweed (*Stellaria media*), annual meadow-grass (*Poa annua*) and fat-hen (*Chenopodium album*). After old grassland, a different type of weed flora can be expected, including buttercups (*Ranunculus* spp.) and rushes (*Juncus* spp.).

Early germinating weeds restrict grass tillering (5) and usually exacerbate clump development of the crop. Of particular concern is chickweed, especially when allowed to form large, smothering patches in mild winters; this can occur with infestations of 50 plants m^{-2} or less. On the other hand, annual meadow-grass tends to occur at much higher densities, often 3000 plants m^{-2}, and causes substantial reductions in the tiller density of sown species (10). These two species, together with taller growing weeds, are especially damaging to clovers.

Generally, the presence of weeds leads to less predictable swards, with infested swards being more difficult to ensile and of lower quality. Although weeds contribute to total yield, they do not match the all-round performance of most sown grasses; in grazed swards their presence hastens the onset of sward deterioration (1).

Cultural control

A dense, vigorously growing crop is the starting point in controlling weeds. Hence the importance of sound crop rotation; rectification of site deficiencies (by draining, liming etc.); burial of surface trash and weed seeds; creation of a fine, firm, fertile and moist seed bed; sensible choice of well-adapted, vigorous species; timely sowing; use of adequate seed rate (see p. 17).

Undersowing usually results in fewer weed seeds establishing compared with direct sowing, whilst broadcast crops are often cleaner than drilled crops (4).

After sowing, most annual broad-leaved weeds with an upright habit can be checked by topping with a mower – but avoid delayed cutting in the year of sowing. Young grass needs to be encouraged to tiller, so quick, rapid grazing, preferably by sheep, is useful particularly if chickweed is threatening.

Perennial weeds, however, arising from seed or regenerating fragments, cannot be eradicated by grazing and topping. They must be controlled by selective herbicides.

Pre-emergence chemical control

Weeds which germinate before the crop emerges can be highly damaging to growth and tillering of ryegrass, but their early removal by pre-emergence herbicides is possible. Because the weeds are young, only modest amounts of herbicides are needed to control them. The following pre-emergence treatments can be used:

3.01. *Pre-emergence spraying of* **paraquat** *0.6 to 0.8 kg/ha to control seedling weeds and cereal stubbles before drilling grass*

Allow 4 to 7 days between spraying and drilling for weeds to die. The higher rate is needed for established weeds.

3.02. *Pre-emergence spraying of* **ethofumesate** *at 1.4 kg ha^{-1} to control annual grass weeds, volunteer cereals and common chickweed in autumn-sown ryegrass and tall fescue leys*

Application should be made to moist soil from mid-August to early October and within two days of drilling. (Broadcast crops should be covered with soil, by harrowing and rolling, before ethofumesate is applied.) This treatment is not suitable for swards reseeded without ploughing or leys containing clovers.

3.03. *Pre-emergence spraying of* **methabenzthiazuron** *1.2 kg ha^{-1} to control meadow-grasses, black-grass and chickweed establishing in perennial ryegrass*

This treatment can be used either (a) in the spring on perennial ryegrass undersown in barley or wheat (as soon as possible after drilling both cover crop and grass), or (b) in the autumn with crops drilled before mid-October. Care should be taken to avoid spray overlapping. Do not spray when the crop is under stress from drought, frost, etc. This treatment is not clover-safe.

3.04. *Pre-emergence spraying of* **bromoxynil** + **ioxynil** *in low volume, to control seedling broad-leaved weeds in all-grass mixtures*

Undersown crops may also be treated provided the cereal is at least at the 3-leaf stage, and the undersown seeds have either not been sown or not emerged. The cereal crop should not be rolled within 7 days of application.

Growth stage for post-emergence spraying

Crop growth stage. Herbicides should not be applied until *all* the sown species (including cereals when seeds are undersown) have reached a stage of growth tolerant to the dose to be used (12).

In most cases, grass seedlings need to be at, or beyond, the 2- to 3-leaf stage (Fig. 3.1). At this stage, young grass seedlings will usually tolerate the amounts of herbicides given in Table 3.1

Tolerance of grass seedlings to herbicides increases at the tillering stage, which generally occurs 4–7 weeks after sowing, depending on species and growing conditions.

The development of legume seedlings is described in terms of expanded leaves (Fig. 3.2). On emergence, two small round or oval cotyledons are apparent. They are followed by a single leaf which is the unifoliate leaf or "spade" leaf. All subsequent leaves of clovers have three leaflets; these are called trifoliate leaves.

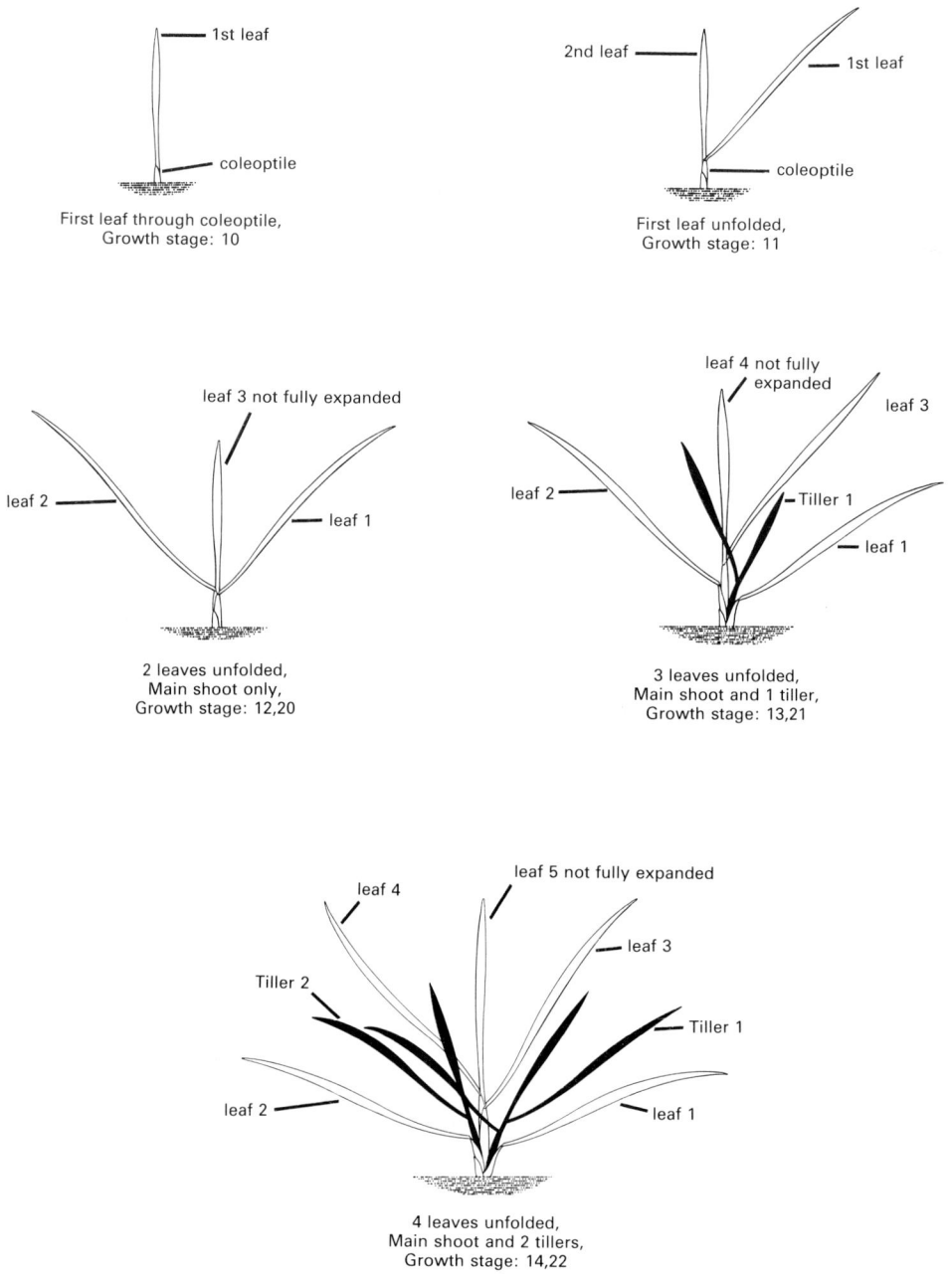

First leaf through coleoptile,
Growth stage: 10

First leaf unfolded,
Growth stage: 11

2 leaves unfolded,
Main shoot only,
Growth stage: 12,20

3 leaves unfolded,
Main shoot and 1 tiller,
Growth stage: 13,21

4 leaves unfolded,
Main shoot and 2 tillers,
Growth stage: 14,22

Fig. 3.1   Growth stages of grass seedlings.

TABLE 3.1. Amounts of herbicides tolerated by perennial ryegrass at the 3-leaf stage

MCPA salt	1.4 kg ha^{-1}
MCPB salt	2.2 „
2,4-D ester	0.5 „
2,4-D amine	0.8 „
2,4-DB salt	2.2 „
Benazolin salt	0.2 „
Dinoseb amine	1.1 „
Mecoprop salt	1.4 „

N.B. Even when grasses and clovers have reached a safe growth stage they may still suffer a serious set-back if they are unhealthy or suffering from pest or disease attack.

Weed growth stage. The younger an annual weed is, the more susceptible it is likely to be to a selective herbicide. Perennial weeds, however, are generally most susceptible when there is a large area of foliage to spray, and when running to flower bud formation. Weeds are often present over a wide range of growth stages at the same time and this may necessitate repeat treatments. Pre-treatment mowing, so that the weeds regrow evenly, can improve the effectiveness of spraying. In general, lush weeds and those growing quickly are more susceptible than those growing slowly.

Controlling broad-leaved weeds in newly sown leys

Where there is a wide variety of broad-leaved weed species present (see Appendix 1 for identification), any of the herbicides listed in Table 3.2 can be used.

Selection of herbicides will be influenced by the weed spectrum (see Table 3.3).

For certain weeds, specific herbicides are needed (see Table 3.4).

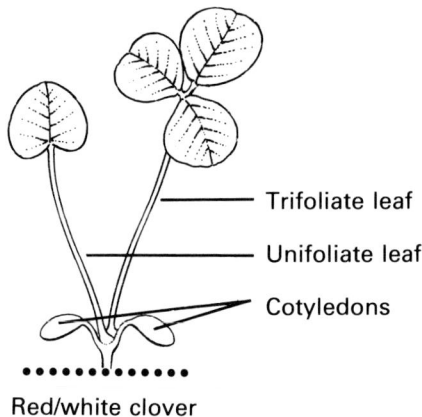

Trifoliate leaf

Unifoliate leaf

Cotyledons

Red/white clover

Fig. 3.2   Clover seedling at the first trifoliate leaf stage.

TABLE 3.2. Herbicides for killing broad-leaved weeds (3–4 leaf stage) in newly sown (a) grass/clover, or (b) all-grass, leys

(a) Grass/clover leys Herbicides	(b) All-grass leys As above, plus
1. Benazolin + 2,4-DB + MCPA	11. Bentazone + dichlorprop
2. Bentazone + MCPB + MCPA	12. Bromoxynil + ioxynil
3. 2,4-DB	13. Bromoxynil + ioxynil + mecoprop
4. 2,4-DB + MCPA	14. 2,4-D
5. 2,4-DB + 2,4-D + MCPA	15. Dicamba + dichlorprop + MCPA
6. Dinoseb acetate	16. Dicamba + mecoprop + MCPA
7. Dinoseb amine	17. Linuron
8. MCPB	18. MCPA
9. MCPB + MCPA	19. Mecoprop
10. Linuraon + 2,4-DB + MCPA	20. 2,3,6 TBA + dicamba + MCPA + mecoprop

Note: All these herbicides (except 2,4-D) can be used on crops undersown in cereals. (The herbicides have been grouped for their safety towards clovers.)

TABLE 3.3. Susceptibility to herbicides of broad-leaved weed seedlings in newly sown grass/clover leys

Herbicide:*	1	2	3	4	5	6	7	8	9	10	11	12	13	14	15	16	17	18	19	20
Black bindweed	S	S	MR	MS	MS	S	MS	R	MR	—	S	S	S	R	S	S	—	R	R	S
Buttercup spp.	S	—	S	S	S	—	R	S	S	—	S	—	S	MS	S	S	—	S	MS	S
Charlock	S	S	MS	S	S	S	S	MS	S	—	S	S	S	S	S	S	—	S	S	S
Chickweed, common	S	S	R	R	—	S	S	R	R	—	S	S	S	R	S	S	S	R	S	S
Cleavers	S	S	R	R	—	MS	MR	R	R	—	S	—	S	R	MS	S	—	R	S	S
Dead-nettle	R	R	R	R	—	—	—	R	R	—	MR	S	S	R	R	R	—	R	R	R
Dock	MS	—	MS	MS	MS	—	—	MS	MS	—	MR	—	MR	MS	MS	—	MS	—	S	S
Fat-hen	S	S	S	S	S	S	MS	S	S	—	S	S	S	S	S	S	—	S	S	S
Fumitory, common	S	S	MR	MS	MS	S	MS	MR	MS	—	S	S	S	R	MS	S	—	MR	MS	S
Groundsel	MS	S	R	MR	MS	S	MR	R	R	—	S	S	S	R	MS	S	—	R	MR	MS
Hemp-nettle, common	S	S	R	MS	MS	S	S	MR	MS	—	MR	—	MS	R	MS	MS	—	S	MR	S
Knot-grass	S	S	MR	MS	MS	MR	MR	R	MR	—	MS	S	S	R	S	S	—	R	R	S
Marigold, corn	R	S	R	R	—	MR	MR	R	R	—	S	S	MS	R	R	R	—	R	R	—
Mayweed, scentless	S	S	R	R	—	MR	MR	R	R	—	S	S	S	R	MS	S	—	R	MR	S
Nettles, small	S	S	MS	MS	S	S	MR	MS	S	—	S	S	S	MS	S	S	—	MS	S	S
Pansy, field	R	MS	R	R	—	—	MR	R	R	—	R	—	MS	R	R	R	—	R	R	R
Parsley-piert	—	—	R	R	—	S	MS	R	R	—	—	—	MS	R	R	R	—	R	R	R
Penny-cress, field	S	S	MS	S	S	S	S	MS	S	—	S	S	S	S	S	S	—	S	MS	S
Poppy, common	S	MS	MR	MS	MS	S	MR	MR	MS	—	MS	S	S	MR	S	S	—	MS	MR	S
Radish, wild	MS	—	R	MS	S	S	MS	R	MR	—	S	S	S	MS	S	S	—	S	S	S
Redshank	S	S	MS	MS	MS	S	MR	R	MR	—	S	S	S	R	S	S	—	R	R	S
Shepherd's-purse	S	S	MS	S	S	S	S	MS	S	—	S	S	S	MS	S	S	—	S	S	S
Sow-thistle, smooth	S	MS	MS	MS	MS	S	MR	MS	MS	—	S	S	—	MR	S	S	—	MS	MR	MS
Speedwell, common	—	R	R	R	—	S	MR	R	R	—	MR	S	S	R	MR	—	—	R	MR	S
Speedwell, ivy-leaved	—	R	R	R	—	S	MS	R	R	—	MR	S	S	R	MR	—	—	R	MR	S
Spurrey, corn	R	S	R	R	—	S	MR	R	R	—	S	—	S	R	S	S	—	R	R	S
Thistle, creeping	MS	MS	MS	MS	MS	—	—	MS	MS	—	MR	—	MR	MS	MS	MS	—	MS	MS	MS

*Herbicides numbered as in Table 3.2.

Key: S = acceptable control   MR = temporary suppression
      MS = effective suppression R = no useful effect

TABLE 3.4. Herbicides for controlling certain specific weeds in newly sown leys (including those undersown in cereals)

Weed species	Appropriate herbicide
Corn marigold (*Chrysanthemum segetum*)	Bentazone mixtures, bromoxynil mixtures, dinoseb
Hemp-nettles (*Galeopsis* spp.)	MCPB mixtures
Cleavers (*Galium aparine*)	Benazolin mixtures, bentazone mixtures, dinoseb
Mayweeds (*Matricaria* spp.)	Bentazone mixtures, bromoxynil mixtures, dinoseb
Knotgrass (*Polygonum aviculare*)	Benazolin mixtures, bromoxynil mixtures, 2,4-DB and mixtures
Black-bindweed (*P. convolvulus*)	Ditto
Redshank (*P. persicaria*)	Ditto
Common chickweed (*Stellaria media*)	Benazolin mixtures, bentazone mixtures, dinoseb

Controlling chickweed in newly sown leys

Common chickweed (*Stellaria media*) is one of the most frequent weeds in young leys, especially those sown in the late summer; a recent survey showed that nearly two-thirds of such sowings were seriously contaminated by this species (4).

Chickweed seed germinates mainly in the spring and autumn. Growth is favoured by moist, fertile soils, mild autumns and thin crops. It can spread from quite modest plant populations to form dense clumps which smother the sown species, reducing grass tillering and clover establishment. Heavy infestations of chickweed can be particularly troublesome in crops grown for conservation – impeding mechanical harvesting and reducing silage quality. There is also evidence that it can cause digestive upsets in ruminant animals.

Choice of herbicide will be dictated largely by the range of other weeds present, time of year and whether or not clover is required. For instance, treatments 3.05 to 3.07 (below) cannot be used on newly sown leys containing clovers (8).

3.05. **Ethofumesate** *at 1 kg ha^{-1} for controlling chickweed in direct-sown ryegrasses and tall fescue leys*

Best results are obtained if the crop is growing vigorously in moist soil and when rain falls within 10 days of spraying. The crop must have at least two or three leaves and can be sprayed from mid-October onwards – but before rapid spring growth has started. This treatment gives residual control of germinating weeds for several weeks.

Where chickweed infestation is heavy and extended a second application may be required: the first application should be made either pre-emergence or early post-emergence in autumn between mid-October and the end of December, followed by the second application no more than 3 months later. This treatment will give additional suppression of annual weed grasses (see 3.16).

3.06. **Linuron** *at 0.25 kg ha^{-1} for controlling chickweed in direct- and undersown, all-grass leys*
This treatment can be applied up to November to all grass species. Do not allow stock to graze for a period of 5 months after application.

3.07. **Mecoprop** *at 1.4 to 2.8 kg ha^{-1} for controlling chickweed in direct-sown all-grass leys*
Spraying can be carried out during frost-free weather in the winter. The higher rate is needed on well-established chickweed, provided grasses are well tillered.

3.08. **Benazolin + 2,4-DB and MCPA** *for controlling chickweed in direct- and undersown grass/clover leys*
Spray when most clovers have reached the one trifoliate leaf stage, before the end of August. Do not roll, harrow or graze within 3 days of spraying. Avoid spraying in drought, excessive rain or frost.

3.09. **Bentazone + MCPB + MCPA** *for controlling chickweed in direct- and undersown grass/clover leys*
Apply when all clover has reached the one trifoliate leaf stage. Grasses should be tillering and chickweed should be at the seedling stage. Do not treat after the end of September.

3.10. **Dinoseb acetate** *at 2.8 kg ha^{-1} for controlling chickweed in direct- and undersown grass/clover leys*
Clovers should have at least one to three trifoliate leaves and grass three or four leaves when treated.

3.11. **Dinoseb amine** *at 2.3 kg ha^{-1} for controlling chickweed in direct- and undersown grass/clover leys*
The clovers should have at least two trifoliate leaves and the grass at least four leaves when sprayed. It is a contact weedkiller and should be applied at high volume. Do not spray when crop is wet, or when rain or frost is imminent.

3.12. **2,4-DB + MCPA** *for controlling chickweed in direct- and undersown grass/clover leys*
Spray when the majority of clover seedlings have one trifoliate leaf. Can be applied at any time of the year, except during drought, heavy rain, waterlogging, frost or extremes of temperature. Weed kill may be slower in cold weather. Some scorch or suppression of clovers may be caused.

3.13. **Linuron + 2,4-DB + MCPA** *for controlling chickweed in direct- and undersown grass/clover leys*
Spraying can be carried out at any time after the two-leaf stage for grasses and one trifoliate leaf stage for clovers, except during drought, waterlogging, frost, heavy rain and extremes of temperature: avoid damage by vapour drift.
f15h9l10m30

Controlling grass weeds, wild oats and volunteer cereals in newly sown leys
(See Appendix I for identification key)

Annual meadow grass (*Poa annua*). This grass, often with rough meadow grass (*P. trivialis*), is one of the most frequent invaders of newly sown leys. It can cause marked reduction in crop tillering and impair the productive life of the sward (6).

Cultural methods of control revolve around producing a dense, vigorously growing crop; infestations of meadow grasses are usually less when seedbeds are ploughed rather than rotovated (9).

For chemical control, early spraying is preferable to later spraying, when grass weeds are more difficult to kill. Pre-emergence spraying is possible with ethofumesate (see 3.02) and methabenzthiazuron (see 3.03).

3.14   *Post-emergence application of* **ethofumesate** *at 2 kg ha^{-1} for controlling annual meadow grass in all-grass leys*

See 3.05 for details.

3.15.   *Sequential applications of* **ethofumesate** *at 1 kg ha^{-1} for controlling annual meadow grass in all-grass leys*

The interval between applications should not exceed 3 months (see 3.05 for details).

3.16.   **Methabenzthiazuron** *at 3 kg ha^{-1} for controlling meadow grasses in perennial ryegrass*

This treatment is recommended for use in the autumn after the removal of the wheat or barley cover crop, so long as the crop is not under stress from drought, frost, etc. (See also 3.03 for pre-emergence application.)

Wild oats. These can be very competitive to newly sown grasses and clovers, as well as posing a threat to subsequent arable crops.

The herbicides listed in Table 3.5 can be used to control wild oats in leys undersown in cereal crops.

TABLE 3.5.   Wild-oat herbicides for use on cereals under-sown with grasses

Herbicide
Benzoylprop-ethyl
Diclofop-methyl*
Difenzoquat
L-flamprop-isopropyl
Flamprop-methyl

*Not clover-safe

25

3.17. **Difenzoquat** *at 1 kg ha⁻¹ for controlling wild oats in ryegrass seed crops*

Apply in autumn, after the three-leaf stage of the crop. A second application may be necessary in the spring.

Volunteer cereals. Modest infestations can be removed by cutting or grazing. Heavier infestations can be controlled by spraying.

3.18 **[For information]** TCA at 5 kg ha⁻¹ can control volunteer cereals in perennial ryegrass

The crop must have at least three leaves and there may be some check to growth. Heavy infestations of cereals and various weed grasses can be tackled by using TCA as above, followed at least 3 weeks later by 1.4 kg ha⁻¹ ethofumesate.

Pests of Establishing Swards

The most important pests of establishing grass are frit fly, leatherjackets and slugs. Recent work suggests that nematodes may cause losses to establishing grass seedlings on light sandy soils. Other pests e.g. wireworms (*Agriotes*) may occasionally cause damage to newly sown grass from time to time, but are generally less important.

Frit fly. Frit fly (e.g. *Oscinella frit*) can cause the total failure of a newly sown sward and although this happens rarely, damage by frit fly often results in patchiness of a recently sown area. Ryegrasses are the worst affected and the risk of damage is greatest in autumn, especially on grass sown after grass and on direct drilled areas.

The term 'frit fly' is used loosely to indicate any small fly larva tunnelling within the base of the stems of grasses and includes not only *Oscinella frit* but about 14 other species which together form a species complex. Many of the species are indistinguishable one from another by the naked eye, either as larvae or as adults. The larvae of all common species are whitish, legless and do not exceed 4–5 mm in length. Adult frit flies are small (4 mm) shiny black flies. Adults of some species, e.g. *Geomyza tripunctata* are brown, slightly larger than those of *O. frit*, and have speckled wings. Individual species within the complex reach their population maxima at different times, but a description of the life cycle of *O. frit* is representative of that for most of the species complex (2). The frit complex is widely distributed thoughout the UK.

Each female lays over 200 eggs during late April and May. The larvae hatch within three days and tunnel into grass tillers, mature in 2–3 weeks, pupate and emerge as adults which lay more eggs during June and early July. The resulting larvae again hatch quickly and attack grass tillers. They pupate in late July and August giving rise to a third generation of adults in August and September. The third generation lays eggs and the larvae hatching from them feed until the onset of cold weather, overwintering as mature larvae which pupate the following spring. There may be only two generations per year in the far north and four in the south instead of the more usual three. In practice, however, the generations tend to overlap and since the generation times of species within the complex differ, larvae are present virtually throughout the year, and adults of one species or another can be found from April to September or even October.

Control is simple. A delay in sowing of six weeks or more after ploughing-in an old sward or desiccating it with herbicides before re-seeding allows time for the larvae to die and prevents their invading the newly sown area. An alternative control at establishment is the use of pesticides (chlorpyrifos, cypermethrin or omethoate). Damage to spring sowings is usually slight and control is not required.

Leatherjackets. The larvae of leatherjackets (*Tipula* spp.) graze on roots and can cause significant damage to newly sown swards. Leatherjackets are widely distributed in the UK but are especially numerous in higher rainfall areas and moist patches in fields (3). Populations fluctuate greatly from year to year, but are monitored and forecast by ADAS.

Leatherjackets are the larvae of crane flies or daddy-long-legs. Each female crane fly lays about 300 eggs, mostly from mid-August to early September. Eggs are often laid in or near tufts of longer grass and mowing or grazing these off at the appropriate time may reduce the numbers deposited. Eggs hatch about two weeks after being laid. The larvae, which are very sensitive to desiccation and sunlight, feed in autumn, reaching about 10 mm in length by November. The larvae grow only slowly in winter but feed voraciously in spring, reaching 40 mm long by April. They have no distinct head capsule and are legless, usually a dull greenish-grey colour, feel rubbery and are difficult to squash between thumb and forefinger. Grass, clover and other forage legume plants are attacked by the larvae. Leatherjackets pupate in July and the adults emerge from August to early October.

Insecticides, either chlorpyrifos applied as a high volume drench or gamma-HCH and bran mixtures give effective control, especially if applied during mild humid weather in spring. Surface baits may give better results on young leys.

Slugs. Slugs (e.g. *Deroceras reticulatum* = *Agriolimax reticulatum*), especially the grey field slug (*D. reticulatum*) can cause extensive losses in grassland, particularly to newly sown areas on heavy soils during spells of wet, warm weather. The grey field slug is however rare on peat soils and dry, sandy soils (3). Eggs are laid mostly in August and the autumn, although breeding can take place at other times. When hatched the young slugs resemble adults in general appearance but are of course much smaller. Slugs reach maturity 12–15 months after hatching. Damage is more likely to occur in swards established by direct drilling than those sown after ploughing and cultivating. Consolidation of the seed bed restricts movement by slugs, and consequently their ability to cause damage. Producing a fine tilth also helps by reducing the number of cracks and crevices in which slugs can lurk. Control can be achieved by the use of metaldehyde or methiocarb pellets.

Eelworms (nematodes). Although grassland normally supports and can tolerate large numbers of eelworms, recent evidence suggests that there are some situations where nematode damage is more likely to occur, e.g. on light, sandy soils especially during the period immediately following sowing. The order Nematoda is a large one comprising mostly very small vermiform animals, few of which are easily visible to the naked eye. They exist in large numbers in soil under grassland and some, e.g. *Pratylenchus, Tylenchorhynchus, Xiphenema*, are parasitic on plant roots (14). Some, e.g. *Meloidogyne naasi*, cause galls or cysts to form on roots. The threshold population level at which the many parasitic eelworm species cause damage to grassland was thought to be very high. A few species, e.g. *Ditylenchus*, live within plant tissues and may cause great damage to legumes.

Unfortunately nematodes are difficult to control in practice. Pesticides which reduce their populations also kill beneficial species such as earthworms and consequently no nematicides are approved for use on grassland. Alternative control measures for use on grasses have not yet been developed.

Wireworms. Sometimes seed may be eaten by wireworms (*Agriotes* spp.) and seedlings are often bitten through at or below the soil surface, resulting in a patchy establishment. Ploughing, and the cultivations involved in preparing a seed-bed for re-seeding, greatly reduce wireworm populations, but enough are often left to impair the newly sown sward.

Although they cause damage to a range of crops, wireworms are denizens of grassland (7). They are the larvae of click-beetles and have a long life cycle (3–5 years). They are yellow, shiny and have short legs and feel hard to the touch. Mature larvae reach 20–25 mm in length. Each female beetle lays 40–150 eggs during the period April–June, which hatch after 5–6 weeks. At first the larvae feed mostly on decomposed vegetable matter, but subsequently on living plant roots and underground stems. Feeding activity is greatest during March to May but continues until September. Larvae migrate vertically in the soil, and during winter many will be found as deep as 60 cm. They take 3–4 or even 5 years to mature and pupate in late July. The pupal period is about 3–4 weeks. Adults emerge in August/September and overwinter either in the soil or in tufts of grass, becoming active again in April when egg-laying commences.

Wireworms are distributed throughout the UK, but are most numerous in the south and east; they are markedly less common in the north. Their population in all regions is declining, largely as a result of the use of organochlorine soil insecticides. They can tolerate large differences in soil nutrient status and pH but fewer are found in more acid and in lighter soils. Although more common in the drier regions of the UK, the larvae are susceptible to desiccation and more are found in fields which are relatively wet in summer than in those which are droughty.

The risk of damage by wireworms to establishing swards can be lessened by cultural practices. Ensuring a firm seed bed restricts movement by the larvae and damage is least when there is a well-consolidated seed-bed below a good tilth. Ploughing and sowing in early autumn, when the larvae are less active, also reduces the risk of wireworm damage. Sowing immediately after cultivations are completed, rather than leaving a break, enables the larvae to continue feeding on old buried turf rather than solely on the newly sown seedlings. Chemical control in grassland by the use of the organochlorine compounds aldrin, dieldrin and DDT is no longer permitted.

Seedling and Establishment Diseases

A composite of fungal pathogens is responsible for seedling death, both pre- and post-emergence. Good establishment is vital, since failure at this stage cannot afterwards be rectified except by reseeding, costly in time and cash. Even partial establishment is likely to lead at best to sward imbalance, and at worst to weed encroachment.

Pre-emergence damping off, in which the deterioration process begins soon after the seed coat is broken and germination usually proceeds little further, is caused mainly by species of *Pythium*, and less commonly by *Fusarium*, *Drechslera* and *Cylindrocarpon* spp. These fungi are commonly seed borne. Post-emergence damping off affects seedlings from the first or second leaf stage onwards, causing death through rotting of the root and stem base, lesions on the coarser roots, and invasion of the laterals. The fungi listed above are also responsible for post-emergence death, along with *Rhizoctonia solani*, *Cladochytrium* spp. and other root-invading fungi.

A well-drained seed bed, together with adequate phosphate and nitrogen, will do much to lessen these problems. Seed dressing is now quite common practice. A liquid seed treatment with drazoxolon (150 ml 100 kg^{-1}) has proved highly efficaceous in controlling damping off and has considerable advantages over the organo-mercurial compounds found to be partially successful by Michail and Carr (11), not least from the safety aspect.

References

1. Charles, A. H.; Haggar, R. J. (Eds.) (1979). *Changes in sward composition and productivity.* Occasional Symposium No. 10, British Grassland Society, 253 pp.
2. Clements, R. O. (1980). The importance of frit fly in grassland. *ADAS Quarterly Review,* **36,** 14–26.
3. Edwards, C. A.; Heath, G. W. (1964). *The principles of agricultural entomology.* London: Chapman & Hall, 418 pp.
4. Haggar, R. J. (1979). Grass reseeding survey, 1977–78. *Miscellaneous Publication, British Grassland Society,* 13 pp.
5. Haggar, R. J.; Kirkham, F. W. (1981). Selective herbicides for establishing weed-free grass. *Weed Research,* **21,** 141–151.
6. Haggar, R. J.; Parsons, A. (1978). Some consequences of controlling *Poa annua* in newly sown ryegrass leys. *Proceedings 1978 British Crop Protection Conference – Weeds,* 301–308.
7. Jones, F. G. W.; Jones, M. (1964). *Pests of field crops.* London: Edward Arnold, 406 pp.
8. Kirkham, F. W.; Haggar, R. J. (1983a). Chickweed control in autumn-sown leys. *Technical leaflet No. 28, Weed Research Organization,* pp. 3.
9. Kirkham, F. W.; Haggar, R. J. (1983b). Establishing ryegrass leys free of seedling grasses. *Technical leaflet No. 8, Weed Research Organization,* 2 pp.
10. Kirkham, F. W.; Haggar, R. J.; Elliott, J. G. (1982). Controlling weeds during grass establishment. *9th Annual Report, Weed Research Organization,* 55–61.
11. Michail, S. H.; Carr, A. J. H. (1966). Effect of seed treatment on establishment of grass seedlings. *Plant Pathology,* **15,** 60–64.
12. MAFF. (1983). Weed control in grassland, herbage legumes and grass seed crops 1983–84. *Ministry of Agriculture, Fisheries and Food (Publications) Booklet 2056 (83),* 38 pp.
13. Roberts, H. A. (1982). *Weed Control Handbook: Principles. Chapter 12, Weed control in grassland,* 351–368. Oxford: Blackwell Scientific Publications.
14. Spaull, A. M.; Clements, R. O. (1982). The effects of root ectoparasitic nematodes upon grass establishment. *Grass & Forage Science,* **37,** 183.

Chapter 4   THE MANAGEMENT OF
ESTABLISHED GRASSLAND

Grass is grown to feed livestock. Though practising farmers scarcely need to be reminded of this it is easy for grassland enthusiasts and advisers to concentrate on the production of a large and uniform crop of herbage without sufficient regard for its profitable utilization. This approach may lead to expensive chemical treatment or other measures which simply do not pay for themselves in increased animal output. If grassland is carefully manured and managed, specific weed and pest control measures may often be unnecessary. Certainly if the primary aim is profitability rather than visual appearance the likely financial returns of any control operation must be evaluated as far as possible.

The primary aim, then, in grassland management is to provide a continuous supply of herbage, either fresh or conserved, that is sufficient in quantity and quality to sustain the desired level of animal production. If herbage productivity is inadequate the first area in which to look for improvement is soil fertility. Soil water regime is also important, but drainage or irrigation measures can be very expensive. Next consider frequency and severity of defoliation, under both cutting and grazing. This has a profound influence on quantity and quality of herbage utilized.

Only when these environmental and management factors have been altered will it be worth attempting to change botanical composition by means of chemical treatment. Indeed, it is often found that once fertility has been increased and management intensified, swards become dominated by perennial ryegrass (*Lolium perenne*), with white clover (*Trifolium repens*) at lower N levels. Other grasses and broad-leaved species are suppressed or eliminated because they are less competitive under these conditions. The effect of selective herbicide application to established grassland will usually be short-lived, unless preceded or accompanied by more fundamental improvements.

The Soil

Soil pH and nutrients

The optiumum pH for soils under grassland is reckoned to be 6.0. Calcareous soils, which are more alkaline than this, do not usually depress grass growth markedly, but in the acidic conditions which naturally obtain in many British soils, particularly in higher rainfall areas, important nutrients may not be available to plants and those species which are tolerant of acid conditions will dominate the swards. When pH is between 4.5 and 5.5 swards are often comprised largely of bent grasses (*Agrostis* spp.) and Yorkshire-fog (*Holcus lanatus*). These species have a slightly lower potential yield than perennial ryegrass, although under these more adverse conditions they may be equally or even more productive. On very acid soils, normally found on heath or moorland, species of very low productivity such as mat grass (*Nardus stricta*) and purple moor-grass (*Molinia caerulea*) predominate. The presence of common sorrel (*Rumex acetosa*) can be a useful indicator of acid conditions. Liming to achieve a pH approaching 6.0 is invariably worthwhile financially if increased productivity is required, and is essential if applications of N, P and K are to be fully exploited. It is particularly important if legumes are to be encouraged.

Phosphate and potash

These are both essential for plant growth, and many soils need regular applications of P and K fertilizer in order to sustain acceptable herbage yields. This is particularly important if there is no return of nutrients via dung and urine, i.e. on cutting fields, especially those not receiving farmyard manure.

The best widely available guide to requirements is soil analysis. Nutrient status is usually expressed by using indices; soils with indices 0 or 1 require fairly liberal dressings of fertilizer. Soils with index 2 or above may only require substantial amounts if large cuts are being removed (see MAFF booklet 2044). It is essential to avoid excessive build-up of, for example, potash in continuously grazed swards as this can affect magnesium levels adversely. Herbage analysis can be useful when a problem is suspected.

Swards on soils with deficiencies of P and K may well be dominated by common bent (*Agrostis capillaris*) or red fescue (*Festuca rubra*); *Lolium perenne* and other sown species do not thrive under these conditions. However, it is unwise to use botanical composition as indicative of P and K status, since the sward is also greatly influenced by N and defoliation management.

Nitrogen

Application of N fertilizer, balanced if necessary by P and K, is by far the most powerful tool for increasing herbage yield. On almost all soils an economic response can be achieved with applications of up to 300 kg ha^{-1}, and on sites with a good supply of water applications of 450 kg ha^{-1} or more may be worthwhile. This, of course, presupposes that the stocking rate is high enough to utilize the extra grass, and does not lead to major sward damage.

High N favours perennial ryegrass and this species frequently comprises 70% or more of the ground cover in intensively managed swards. *Poa trivialis* is also often present in such swards, but seldom contributes more than 20%; this is very unlikely to affect productivity significantly. Couch (*Elymus (Agropyron) repens*) occasionally invades high N swards, particularly in drier eastern areas. Although high-yielding this grass is not very palatable and is undesirable in an arable rotation. However, once established in a grass sward it is very difficult to eradicate except by complete sward destruction. Frequent grazing is probably the best means of control. Docks (*Rumex* spp.) are also found under high N conditions, but should not be regarded as entirely worthless, since they may well be eaten by cattle, particularly when ensiled (see also next Chapter).

Under lower N conditions swards are likely to be more diverse; it is important to recognize that unsown grasses such as Yorkshire-fog (*Holcus lanatus*) and common bent (*Agrostis capillaris*) are not inherently low yielding and should be regarded as indicators of possible N deficiency rather than causes of low productivity. This may often be the case during the 'lean years' in the third and fourth harvest year after establishment.

Farmyard manure and slurry

Organic manures are potentially valuable sources of plant nutrients, and may also improve soil structure. One of the main problems with them is the spread of weed seeds. Docks (*Rumex* spp.) are particularly associated with applications of slurry on dairy farms. More frequent defoliation is probably desirable, in order to prevent seeding and the formation of large plants which are difficult to kill. In severe cases there is little alternative to the use of selective herbicides. Another potential problem is sward damage from heavy manure applications. This can lead to bare patches and subsequent re-colonization by low-yielding annual meadow grass (*Poa annua*) and chickweed (*Stellaria media*).

Soil Physical Conditions

Grass growth requires that the plant roots are adequately supplied with water and air, and the most productive species tend to have the most exacting requirements.

Shortage of water

Under very dry conditions swards can thin out and allow annual species such as annual meadow grass, wall barley (*Hordeum murinum*) and chickweed to develop. Irrigation will obviously increase production under these conditions, and help to maintain dense swards. Though not widely practised on grassland it can be profitable provided expensive water storage reservoirs are not required (see MAFF Grassland practice No. 16). The avoidance of very close cutting, particularly of a heavy conservation crop, will help to reduce sward thinning due to drought.

Excess of water

Under very wet conditions, usually caused by low soil permeability or high rainfall, soil aeration can be greatly reduced. Soil structure is usually poor and root development may be weak in water-logged soils, and plant diseases are encouraged. More important than the limitation on grass growth, however, is the restriction on utilization of grass on badly drained land. This is particularly important in spring and autumn. If swards are grazed when the soil is too wet to support the weight of the livestock, poaching will occur. The resultant damage, if severe, will reduce herbage production. It may also result in the ingress of *Poa annua* (low yielding) and, more seriously, rushes (*Juncus* spp.) which are unpalatable. Similarly the use of heavy machinery when the soil is wet can result in sward damage. Damage by livestock or machinery can result in loss of structure of the surface soil, or the development of a soil pan (impermeable layer) which further impedes permeability, aeration and root development.

The drainage status of almost all wet soils can be improved. Simple and cheap operations, such as ditch improvement or subsoiling, may be all that is needed; but often a full scale pipe system with moling as a secondary treatment is required. This is expensive and the likely returns should be carefully considered. For many farmers these will derive not primarily from an increase in herbage production, although early spring growth should be increased, but from the ability to fertilize and graze perhaps 2 weeks earlier in spring and 2 weeks later in the autumn without excessive poaching damage. In wet years there may also be improved utilization in the main part of the growing season.

Defoliation

Frequency

The more frequently a sward is defoliated the denser its base will become. Swards which are only cut once a season, at a very mature stage of growth, will yield large quantities of herbage (of low quality), but will become very thin at the base. This allows the development of erect-growing weeds which subsequently run to seed and reproduce.

This problem is much reduced if swards are cut three or four times in the year, particularly at higher N levels. Similarly, swards which are continuously grazed will be denser than rotationally grazed swards and be less prone to weed ingress. They may also offer more resistance to poaching.

Severity

When combined with frequent defoliation, severe grazing, down to say 2 or 3 cm, will tend to encourage prostrate grasses and broad-leaved species capable of creeping, e.g. Buttercups (*Ranunculus* spp.) or rosette plants, such as dandelions (*Taraxacum officinale*) and plantains (*Plantago* spp.). Again, this will be more of a problem when fertility is low.

Selectivity

Given the chance livestock will eat the most nutritious and palatable herbage in a sward, and reject the rest. This will happen at low stocking rates and will encourage species such as creeping thistle (*Cirsium arvense*) and ragwort (*Senecio jacobaea*). At higher levels of stocking animals will be less selective, and in general a more productive sward will result. However, this approach cannot be taken too far, otherwise individual animal performance will suffer. Cattle tend to be less selective than sheep; horses are particularly discriminating. There are therefore advantages in mixed or alternate stocking.

Seasonality

It is wasteful to leave late-summer and autumn grown herbage unused, so a certain amount of autumn and winter grazing is highly desirable; this may also reduce winter kill. Sheep are very useful for this. Severe grazing in winter and early spring, however, particularly if accompanied by poaching damage, is likely to increase weed content. On the other hand, hard grazing in May, during the peak of grass growth, is likely to be beneficial and may encourage white clover.

Effects of the grazing animal

Potential damage from poaching has already been mentioned, and if at all possible stock should be kept off the land when its bearing strength is insufficient to support them. Dense swards are less likely to suffer poaching damage than open swards. The provision of hard tracks can also be useful for minimizing damage. Under dry conditions treading is likely to favour perennial ryegrass.

Dung and urine, although important for the recycling of plant nutrients, can cause sward problems. Areas around cattle dung pats will be rejected by cattle for several weeks, and this may allow tall weeds to develop. Under high N conditions 'urine scorch' can occur – this results in bare patches which are likely to be colonized by annual species such as *Poa annua*.

Topping

This is the practice of cutting grass seed heads and tall weeds on grazing fields, usually in mid-summer. It must be regarded, to some extent, as an admission of failure, and is in many cases largely cosmetic in its effects. Common rush (*Juncus effusus*) is susceptible to regular cutting, other rushes less so. Creeping thistle (*Cirsium arvense*) needs to be cut quite close to the ground and preferably twice per year. Bracken (*Pteridium aquilinum*) should also be cut twice; a single late cut will have little effect. Docks are not weakened by cutting, in fact they are likely to thrive on it if fertility is high. Cutting of ragwort tends to encourage growth in the following year; furthermore the cut plants represent a serious poisoning risk. (See MAFF leaflet No. 280).

White Clover

White clover has great potential as a constituent of long-term grassland. As well as its contribution to soil nitrogen supplies (a grass/clover sward without N fertilizer can produce as much as a pure grass sward receiving more than 200 kg ha^{-1} N) it can increase herbage intake and animal performance, particularly with sheep. At the moment, however, relatively few farmers rely on white clover to any great extent. This is partly because a mixed grass/clover sward is more difficult to manage, and a considerable research effort is being made to improve our knowledge of its requirements.

As has already been mentioned, legumes in general are more sensitive to low pH and low P than are grasses. Basic slag, being a slowly available source of P, has traditionally been used to encourage clover. The more widely available rock phosphates may not be as effective in this respect, as they need slightly acid conditions in order to dissolve.

Nitrogen applications must be low if white clover is to be exploited. A spring dressing of, say, 50 kg ha^{-1} N may not depress clover too much, provided the grass is not allowed to get too long, but generally speaking the more nitrogen is supplied from fertilizer the less will be obtained from clover.

White clover is more sensitive to drought than grasses and shows very good responses to irrigation.

A crucial factor in the use of mixed swards is the competition between the legume and the grass. More frequent defoliation tends to help the clover, which is unlikely to thrive in a conservation regime unless several cuts are taken – 4-weekly intervals may be the optimum. The first cut in spring should be early. It is particularly important to graze autumn and winter growth in mixed swards, otherwise the grass is likely to become dominant. Long-petioled cultivars of clover (e.g. Blanca) tend to perform better than more prostrate varieties when defoliation is infrequent.

When clover is making a significant contribution to the sward it is obviously important to use only clover-safe herbicides. (See MAFF booklet No. 2056.)

Putting Together a System

The main guideline for grazed grass is to match herbage production with requirements, or *vice versa*. If stocking rate is too low, herbage goes to waste and erect-growing weeds will increase. If it is too high sward damage may result, with the ingress of 'opportunist' species; prostrate broad-leaved species will also increase under hard grazing if fertility is low.

Mixed stocking of cattle and sheep (and horses if applicable) is a useful option to consider. The grazing behaviour of livestock of different species tends to be complementary, mainly because of the willingness of one species to graze close to areas dunged by the other.

Conservation policy will depend largely on the quality of winter feed that is required; the higher the quality the more frequently swards need to be cut. It is very difficult to maintain dense, weed-free swards under a traditional system where a very heavy cut is taken in late June or July. It is particularly important to replenish soil K on fields cut repeatedly.

Ideally, the aim should be to alternate cutting and grazing on the same sward in successive years. This will help to prevent the thinning of swards associated with cutting, and the patchiness associated with grazing; it will also help in the recycling of nutrients via dung and urine. Alternation within the season is also possible, though care should be taken to avoid contamination of silage crops by dung and soil. Increasing cutting height may solve this problem.

Finally, it should be recognized that, although there is a shortage of precise information on the effect of weeds on animal production, many unsown species are eaten by livestock and make a contribution to productivity. Indeed, species such as dandelion are of high nutritional quality, although their yield may be low. Therefore the indiscriminate spraying of grassland should be avoided. Chemical treatment should ideally be regarded as a way of reducing large populations of weeds that are obviously unpalatable, unproductive and/or poisonous. Under intensive conditions the likely candidates are docks (although these are in fact quite productive and may be eaten by cattle, particularly when ensiled), chickweed, nettles (*Urtica* spp.) and annual meadow-grass. Under moderately intensive management, particularly with sheep grazing, creeping thistle may need to be controlled chemically, as will ragwort and buttercups. On rough grazings bracken is generally worth spraying.

Management aimed at maintaining a vigorous, weed-free sward will also tend to minimize attacks by pests and diseases. Frequent defoliation is recommended for many of these, while avoidance of poor soil conditions, particularly bad drainage, is a first consideration in prevention or control, particularly of fungal diseases. However, it must be said that conditions which deter some pathogens may encourage others: rusts, for example, cause less damage when nitrogen levels are high, whereas with other diseases the reverse is true.

Further reading

Charles, A. H.; Haggar, R. J. (Eds.) (1979). *Changes in sward composition and productivity.* Occasional Symposium No. 10, British Grassland Society, 253 pp.

Holmes, W. (Ed.) (1980). *Grass – its production and utilization.* Oxford: Blackwell, 295 pp.

Ministry of Agriculture, Fisheries and Food. Grassland practice series (regularly updated).

No. 2   Nitrogen for grassland
No. 4   Phosphate and potash for grassland
No. 9   Pasture improvement, including the use of herbicides
No. 10   Grazing systems for dairy cows
No. 11   Grazing management for beef cattle
No. 12   Grazing management for lowland sheep
No. 13   Grassland irrigation

Advisory leaflets for individual weeds:
No. 46   Weed control: docks
No. 280   Weed control: ragwort
No. 51   Weed control: thistles
No. 89   Couch
No. 190   Bracken and its control
No. 433   Weed control: rushes

Chapter 5   CONTROLLING WEEDS IN PERMANENT SWARDS

As emphasized in the previous chapter, weed infestations are usually associated with deficiencies in soil fertility, pasture management or sward utilization. These causes must be corrected before lasting benefits from weed control measures can be achieved. Conversely, maintaining a vigorous sward is the best way of preventing a build-up of weeds (14).

Cultural Control

Once the conditions which cause the weed problem have been rectified, the weeds themselves can be tackled. Pulling or digging out weeds is not recommended except for light infestations that are uneconomic to deal with in any other way. Neither is the practice of cutting flowering stems since, as with ragwort (*Senecio jacobaea*), this may do more harm than good by encouraging prostrate growth and perenniation. Even regular and frequent topping is often ineffective; it will take at least two years of close and regular cutting to eradicate creeping thistle (*Cirsium arvense*). Hence, serious weed problems need to be tackled by spraying with selective herbicides. Even so they will tend to recur if the field management is unchanged.

Chemical Control

Choice of treatment

The wrong herbicide will kill both crop and weed. Where clovers are important, only clover-safe herbicides should be used (11). After identifying the weeds present (Appendix I) suitable herbicides can be selected by referring to appropriate weed susceptibility tables (Tables 5.1 and 5.2). Most herbicides are sold as mixtures, to widen their weed spectrum of activity. The choice of herbicide will also depend on cost, availability and relative ease of application. The product label should be read carefully before spraying.

TABLE 5.1.   Herbicides for controlling broad-leaved weeds in established swards

(a) Swards containing clovers	(b) All-grass swards
1. Asulam	8. Asulam + mecoprop + MCPA
2. Benazolin + 2,4-DB + MCPA	9. 2,4-D
3. 2,4-DB	10. Dicamba + mecoprop
4. 2,4-DB + 2,4-D + MCPA	11. Dicamba + mecoprop + MCPA
5. Dinoseb acetate/amine	12. Dicamba + mecoprop + 2,4,5-T
6. MCPB	13. 3,6-dichloropicolinic acid
7. MCPB + MCPA	14. MCPA
	15. Mecoprop
	16. 2,4,5-T + 2,4-D ± dicamba
	17. 2,3,6-TBA + dicamba + MCPA + mecoprop
	18. Triclopyr
	19. Triclopyr + 3,6-dichloropicolinic acid
	20. Triclopyr + dicamba + mecoprop

TABLE 5.2. Susceptibility to herbicides of established broad-leaved weeds in permanent swards

Herbicide:*	1	2	3	4	5	6	7	8	9	10	11	12	13	14	15	16	17	18
Bracken	S	—	R	—	—	R	—	—	R	—	—	MR	—	R	R	—	—	—
Buttercup, bulbous	—	R	MR	—	—	MR	MR	—	MR	MR	MR	MR	—	MR	MR	MR	MR	MR
Buttercup, creeping	—	S	S	S	—	MS	MS	S	S	S	S	S	—	S	S	MS	S	MR
Buttercup, meadow	—	—	MS	S	—	MS	—	—	MS	—	—	S	—	MS	MS	S	—	MR
Cat's ear	—	—	—	—	—	—	—	—	MS	—	—	—	—	MS	MS	—	—	—
Chickweed	—	S	R	—	S	R	—	S	MR	S	S	—	—	MR	S	—	S	—
Daisy	—	R	MR	—	—	MR	MS	S	MS	S	S	S	—	MS	MS	S	—	—
Dandelion	—	R	MR	—	—	MR	R	MS	MS	MS	MS	S	S	MR	MR	S	—	—
Dock, broad	S	MS	R	—	—	R	MR	S	MR	S	S	S	—	MR	MS	MS	MR	S
Dock, curled	S	MS	MS	MS	—	MR	MS	S	MS	S	S	S	—	MS	MR	MS	MR	S
Hawkbit	—	—	MS	—	—	MS	MS	—	S	—	MS	—	—	MS	MS	—	—	—
Horsetails	—	MS	MR	MS	—	MR	MR	—	MR	MS	MS	MR	—	MR	R	MR	MS	—
Knapweed	—	—	—	—	—	—	—	—	MS	—	—	—	—	MS	—	—	—	—
Meadowsweet	—	—	R	—	—	R	—	—	MR	—	—	—	—	MR	—	—	—	—
Nettle	—	MS	MS	—	—	MR	MS	MS	MS	MR	MS	S	—	MS	MS	S	MS	S
Plantains	—	S	S	S	—	S	S	S	S	S	S	S	—	S	S	S	—	—
Ragwort	—	R	R	—	—	R	R	MS	MS	MR	R	MS	—	MS	—	MS	—	R
Rush, compact	—	—	—	—	—	MS	—	—	MS	—	—	—	—	MS	—	—	—	—
Rush, hard	—	—	R	—	—	R	—	—	MR	—	—	—	—	MR	—	—	—	—
Rush, heath	—	—	R	—	—	R	—	—	R	—	—	MR	—	R	—	—	—	—
Rush, soft	—	R	MR	—	—	MR	MS	—	MS	MS	MS	MR	—	MS	MR	—	—	—
Selfheal	—	—	—	—	—	—	—	—	MS	—	—	MR	—	MS	—	—	—	—
Sorrel, common	—	—	—	—	—	MR	—	—	MR	S	S	S	—	MR	R	S	—	—
Sorrel, sheep's	—	—	MR	—	—	MR	—	—	MR	—	—	S	S	MR	R	S	—	—
Sow-thistle	—	MS	—	MS	—	—	MR	—	MR	MS	MS	MS	MS	MR	MR	MS	MS	—
Thistle, creeping	—	MS	MS	MS	—	MS	MS	MS	MS	MS	MS	MS	S	MS	MS	MS	MS	MS
Thistle, spear	—	—	MS	MS	—	MS	MS	MS	MS	MS	MS	MS	S	MS	MS	MS	—	—
Wild onion	—	MR	—	—	—	—	—	—	MR	—	—	MR	—	MR	MR	MR	—	—
Yarrow	—	MR	R	—	—	R	—	—	MR	—	—	S	—	MR	MR	S	—	—
Yellow rattle	—	MR	—	—	—	—	—	—	MR	—	—	—	—	MR	S	—	—	—

*Herbicides numbered as in Table 5.1.
Key: S = acceptable control MR = temporary suppression
      MS = effective suppression R = no useful effect

When to spray

For best results spray when weeds are in seedling stage. If this is not possible, spray during periods of active growth but before flowering. With most weeds, this means spray in May or early June, but there are exceptions, e.g. bulbous buttercup (*Ranunculus bulbosus*) and chickweed (*Stellaria media*) in autumn, bracken (*Pteridium aquilinum*) in early August.

Spray times can be modified by grassland management. For instance, grazed fields containing ragwort (*Senecio jacobaea*) should be treated in early May whereas fields for hay should be treated in the previous autumn if this weed is present. Similarly, spraying nettles (*Urtica* spp.) and docks (*Rumex* spp.) after June or July is effective so long as there is adequate regrowth of leaves.

Heavy rain falling within about 6 hours of spraying a translocated herbicide will decrease its effect, and a longer period free from rain is preferable with contact herbicides, e.g. dinoseb and ioxynil. Drought and very hot weather may reduce herbicide activity.

Application

Where weeds are dispersed widely throughout the sward, overall spraying is needed, but avoid spray drift by not spraying in windy weather and by ensuring that the herbicide is applied at the correct pressure. Herbicides that rely on contact action, e.g. dinoseb, ioxynil, should be applied at high volume. Accuracy of application is more important with these herbicides than with growth regulator herbicides. Avoid over- or underlapping and ensure safety standards (to operator, surrounding crops and wildlife) are observed, both during use and in the disposal of any excess herbicide. (See Appendix II.)

For spot treatment of clumps of weeds, a knapsack sprayer is more appropriate. Alternatively, new "weed-wiping" applicators are available for smearing translocated herbicides at high concentrations on to tall growing weeds without damaging the underlying grass and clover.

Management after spraying

When large clumps of weeds have been killed, e.g. nettles (*Urtica* spp.), the bare patches need to be filled by sowing grass or clover seed.

After spraying, fields should be left for at least 7 days to allow the herbicide to be translocated. If poisonous plants are present, special precautions should be taken.

Examples of poisonous weeds include bracken (*Pteridium aquilinum*), buttercups (*Ranunculus* spp.), cowbane (*Cicuta virosa*), deadly nightshade (*Atropa belladonna*), field horsetail (*Equisetum arvense*), foxglove (*Digitalis purpurea*), hemlock (*Conium maculatum*), meadow saffron (*Colchicum autumnale*) and water-dropworts (*Oenanthe* spp.). Of particular concern is common ragwort (*Senecio jacobaea*), especially in hay.

These weeds are not eaten normally when grass supplies are adequate, but they may become more palatable to livestock after spraying or topping. Following spraying, therefore, fields should be left for at least 14 days before being grazed – and much longer if ragwort is present (see p. 45).

Occurrence and Control of Common Weeds of Established Swards

Barley grass/wall barley. Barley grass (*Hordeum murinum*) is an annual species, found in association with perennial pastures weakened by drought, over-grazing or physical disturbance. It is therefore common on stock camps, pathways and under trees in East Anglia. The awned seed affects sheep production directly by becoming entangled in wool, penetrating pelts and blinding lambs.

A combination of herbicide application and changed management practices, including altering fence lines, is most likely to lead to long-term relief from barley grass (13). Continuous hard grazing in spring followed by mowing for silage helps to reduce the production of seed heads. (In New Zealand, the plant growth regulator mefluidide has a label claim for barley grass seed head prevention with a spring application.)

5.01.   **Ethofumesate** *applied at 2 kg ha^{-1} in winter for controlling barley grass in all-grass swards*

This treatment, although expensive, is very effective and therefore useful for limited infestations. Two possible drawbacks are that it has to be applied at a time when it is difficult to see the edges of the barley grass patches, and it takes out clovers. If large gaps occur after treatment of the sward, oversowing may be necessary. Barley grass germinating 12 or more weeks after treatment may not be controlled. Grazing in January or February is not recommended after spraying in October–December, as this is likely to reduce the control of weed grasses. For further details, see 3.15.

5.02. **TCA (63%) + dalapon (11%)** *applied at 7 kg ha^{-1} in late autumn or early spring for controlling barley grass in grass/clover swards*

This treatment is in widespread use in New Zealand. Where large amounts of barley grass or other annual grasses are present, application may need to be followed by oversowing after 8 weeks or in early autumn. Spot applications to heavily infested areas can be made as late as the first signs of flower head appearance.

Bracken. Bracken (*Pteridium aquilinum*) infests an estimated 200,000 ha of hill land in Scotland and about 160,000 ha in England and Wales. It occurs on deep, well drained, fertile soils and its removal can lead to considerable increases in grass growth, depending on the initial density (2).

Losses of cattle and horses, and to a lesser extent sheep, due to bracken poisoning are common, especially when animals are hungry or when they are newly introduced to infested areas. (For symptoms of bracken poisoning see (9)). Other drawbacks include the difficulty of finding and attending sheep in the dense stands, plus the belief that bracken is instrumental in spreading tick-borne diseases in sheep.

Bracken spreads by rhizome extension. Fronds emerge in spring and die off with the onset of autumn frosts. Dead fronds decompose slowly, often forming a deep litter layer.

Where practical, ploughing in mid-summer followed by repeated discing can lead to eradication. Cutting or slashing, preferably at least twice in mid-summer at 6-weekly intervals, can provide a short-term check. Young fronds are easily damaged before they start to open, so rolling or heavy treading at this time can be useful.

5.03. **Asulam** *at 4.4 kg ha^{-1} applied at full-frond expansion stage (normally mid-July to early August) for controlling bracken in grass/clover swards*

Spraying should be carried out in dry weather and when the plants are in full frond, actively growing, bright green and soft to the touch. The fronds should not be damaged by stock, frost or cutting before treatment. Little herbicide effect is seen on the bracken in the year of spraying, although most grasses and some herbs will be severely damaged unless shielded by the bracken. Asulam can be applied to small areas by knapsack, tractor mounted or ultra low volume sprayers; aerial spraying is more appropriate for larger, inaccessible areas, any surviving patches of undamaged bracken being sprayed the following year. Addition of a wetting agent will increase the speed of uptake in adverse climatic conditions. To reduce possible risks of bracken poisoning it is advisable to exclude stock from treated areas until after normal die-back. After spraying, management aimed at preventing regeneration of bracken may include fertilizing, reseeding, fencing and restocking.

5.04. **Glyphosate** *applied at 1.08 kg a.e. ha^{-1} in July and August to kill bracken*

Apply at full frond development. Most grasses and broad-leaved species will also be killed, so reseeding and improved after-care management will be needed. However, to avoid such damage to the underlying vegetation, glyphosate may be applied selectively using a specified height-selective applicator.

Buttercups. The three most common species are bulbous (*Ranunculus bulbosus*), creeping (*R. repens*) and meadow buttercup (*R. acris*). As they differ in their response to herbicides, they need to be identified before treatment. Bulbous buttercup is the first to flower and has swollen stem bases. Creeping buttercup produces runners and has furrowed flower stalks, unlike meadow buttercup.

In the National Farm Study (12), one quarter of swards over 20 years old in England and Wales were found to be infested with buttercups, rising to one third on sites with poor or bad drainage.

All species of buttercups can be poisonous if large amounts are consumed *in the fresh state* at the time of flowering (well dried hay containing buttercups is relatively harmless). Fortunately, their acrid taste prevents them from being eaten in quantity, except by young animals or when choice is limited. Being largely avoided by stock, however, means that they are able to compete with grass under grazing.

Creeping buttercup is reported to be most closely associated with hay crops on poorly drained soils, hence some control of this species is possible by changing management from hay to grazing and improving drainage. This change is less likely to control bulbous buttercup, which is favoured by well-drained soils, especially continuously grazed old swards. In general, buttercups are favoured by horse grazing, but are discouraged by fertilizer nitrogen (5). Mowing to encourage more uniform regrowth at spraying is often helpful, as is a nitrogen application two weeks before spraying, to make the sward more competitive and to discourage fresh weed germination.

5.05. **MCPA salt** *at 1.6 kg ha^{-1}, or* **2,4-D amine** *at 1.4 kg ha^{-1}, or* **2,4-D ester** *at 1.0 kg ha^{-1} in late spring for controlling buttercups in permanent all-grass swards*

Spraying should be carried out just before flowering, during warm weather when the crop is growing actively. Creeping buttercup is the most susceptible of the three species to the above herbicides, whilst bulbous buttercup is the most difficult to kill. Spraying in spring will prevent the latter species from flowering but may not reduce the population in the following year; for this reason it may be necessary to respray in the autumn when new leaves appear and germination occurs. MCPA is usually more effective than 2,4-D for controlling buttercups. After spraying heavy infestations keep livestock out for at least two weeks. For lasting control, repeat sprayings are usually needed. (See also 5.26 for mixture of Mecoprop + asulam + MCPA.)

5.06. **MCPB** *at 2.2 kg ha^{-1}, or* **2,4-DB** *at 2.2 kg ha^{-1} for controlling buttercups in late spring in permanent grass/clover swards*. Apply as per 5.05. These herbicides are safe on white clover.

5.07. **[For information.]** Mixtures of MCPA 0.56 kg ha^{-1}, or 2,4-D amine 0.56 kg ha^{-1}, or 2,4-D ester 0.28 kg ha^{-1}, plus up to 1.4 kg ha^{-1} of either MCPB or 2,4-DB will kill buttercups in permanent grass/clover swards. The rates quoted for MCPA and 2,4-D are the maximum doses that can be applied without causing a substantial check in the legume content of the sward. When red clover is present, 2,4-D should not be used.

5.08. **Mixture of MCPB + MCPA** *for controlling creeping buttercups in grass/clover swards*

Spray before flowering.

5.09. **Mixture of benazolin + 2,4-DB + MCPA** *for controlling creeping buttercup in grass/clover swards*

Spray when weeds are well developed and growing actively but before flowering. Don't spray if frosts, rain or drought are imminent. Bulbous buttercup is unlikely to be controlled satisfactorily.

Common chickweed. Although a major contaminant of newly sown leys (see Chapter 3), chickweed (*Stellaria media*) is also a frequent colonizer of heavily stocked permanent pastures, especially those damaged by autumn poaching or slurry application.

Most of treatments 3.05–3.13 can be used for controlling chickweed in permanent grassland.

Cow parsley. Cow parsley (*Anthriscus sylvestris*) is a biennial plant which is liable to behave as a perennial if prevented from flowering in the second year. It is a frequent component of hay meadows, on fairly infertile soils, and is rarely grazed by livestock.

Cow parsley is resistant to most herbicides.

5.10. **[For information.]** Dichlorprop at 3 kg ha^{-1} has given good control of cow parsley in permanent all-grass swards, when applied in summer at the early flowering stage.

5.11. **[For information.]** Chlorfurecol-methyl at 1.5 kg ha^{-1} plus maleic hydrazide at 3 kg ha^{-1}, applied in spring, has been shown to give good control of cow parsley *but this treatment will severely retard grass growth.*

Soft-brome. This species (*Bromus hordeaceus*) (Plate IIc) can cause serious problems many upland farming systems based on early sheep grazing and hay in northern England and Scotland. Although a high-yielding grass, its hairy leaves and awned seeds make it unattractive to stock. Because it is a rapid and prolific seeder, it normally runs to head before other grasses are ready for cutting. Consequently, masses of new seedlings establish in the autumn.

As with barley grass (see p. 39), long-term control is best achieved by a combination of herbicide treatment and changed management practice (3).

5.12. **Ethofumesate** *at 2 kg ha^{-1} between mid-January and mid-April for controlling soft-brome grass at the 2–3 leaf stage*

5.13. **[For information.]** TCA 2.8 kg ha^{-1} plus dalapon 0.5 kg ha^{-1}, applied in late autumn or early spring, has given good control of creeping soft-grass in permanent swards in northern England. (Note: these rates are lower than those for barley grass control – see 5.02 – because they tend to be rather toxic to the indigenous grasses.)

Docks. Broad-leaved dock (*Rumex obtusifolius*) is the most abundant dock species in UK grassland (7). It has large oval leaves and is often found with curled dock (*R. crispus*), which has long, narrow leaves with wavy margins. Hybrids of these two species, with intermediate characteristics, are also common.

Over 200,000 ha of UK grassland are seriously infested with docks; inclusion of fields with sparser or localized (incipient) infestations gives a value of over 0.5 m ha. Docks are found most frequently on dairy farms, being closely associated with short-term leys cut for silage and the application of slurry, farmyard manure and nitrogen.

Docks have strong tap roots, are profuse seeders, and their seeds can remain viable for many years; if allowed to grow unrestricted they are likely to spread to adjoining arable land. Hence, both species of docks are listed in the 1959 Weeds Act and are scheduled in the 1974 Seeds Regulations for special standards in herbage seed crops.

Docks are coupled with plants having dangerous quantities of nitrates and oxalates – both poisonous to livestock. In silage cutting regimes, infestations exceeding 20% of ground cover can reduce grass growth substantially from July onwards and economic benefits from controlling such infestations can be expected (4). Under grazing systems, docks are often apparently ignored by stock and stand self confessed as non-contributors to productivity. Sometimes, however, the percentage consumption of docks can be surprisingly high, nearly equalling that of grass, yet the lower digestibility of docks makes them only 65% as valuable as associated grass material.

Prevention of seeding, whether in the field or around farm buildings, is clearly the best precaution against spread; feeding of infested hay or spreading contaminated manure on open swards should be specially avoided. Also, practical steps to reduce sward damage by excessive trampling will reduce opportunities for infestations to develop.

Because docks, unlike most other grassland weeds, are not discouraged by intensification of management, herbicidal methods of control have to be adopted.

5.14. **Asulam** *at 1.1 kg ha^{-1} for controlling docks in grass/clover swards*

Plants should be growing actively, be in full leaf and show no signs of flower stem emergence. Hence, before spraying allow plants to recover fully from previous defoliation. After spraying allow 7 days for absorption of herbicide. Repeat application is likely to be required for long-term control, either in late summer or in spring following the first application. Asulam should not be used on hay fields containing substantial quantities of sensitive weed grasses like Yorkshire-fog (*Holcus lanatus*), creeping soft-grass (*H. mollis*), bents (*Agrostis* spp.), bromes (*Bromus* spp.) and meadow grasses (*Poa* spp.).

5.15. **Mixture of benazolin + 2,4-DB + MCPA** *for controlling docks in grass/clover swards*

Apply when shoots begin to form.

5.16. **Glyphosate** *at 1.5 kg a.e. ha^{-1} for killing docks in grass/clover swards by directed application*

This treatment (applied using a specified height selective applicator) will only be selective if there is an adequate height difference between the crop canopy and weed.

5.17. **MCPA** *at 1.7 kg ha^{-1} or* **2,4-D amine** *at 1.4 kg ha^{-1} or* **2,4-D ester** *at 1.0 kg ha^{-1} for controlling seedling docks in all-grass swards*

Peak periods of germination occur in April and September, or after cutting for silage.

5.18. **Mecoprop** *at 2.8 kg ha^{-1} for controlling docks in all-grass swards*

Mature docks may be cut in July, after flowering but before seeding, and the regrowth sprayed one month later. This treatment is likely to need repeating in the following year.

5.19. **Triclopyr** *at 1.4 kg ha^{-1} for controlling docks in all-grass swards*

Do not spray in drought, or when sward or weed is under stress. Control will be reduced if rain falls within 2 hours of application.

5.20. *Mixture of* **triclopyr** *and* **3,6-dichloropicolinic** *acid for controlling docks in all-grass swards*

Treat in the spring when the docks are in the rosette stage, 15–20 cm high. If previously defoliated, allow 2–3 weeks regrowth before spraying. On large, well established docks, or where there is a large reservoir of seed in the soil, a second application the following year may be required.

5.21. *Mixture of* **triclopyr** + **dicamba** *and* **mecoprop** *for controlling docks in all-grass swards*

This treatment is more suitable when other weeds are present, *e.g.* nettles (*Urtica* spp.) and thistles (*Cirsium* spp.). Where docks are growing from well established rootstocks, retreatment may be required in the following year.

5.22. **Mixture of asulam** + **mecoprop** + **MCPA** *for controlling docks in all-grass swards*

Apply as per 5.14. Clovers will be damaged by this treatment; the weed spectrum is wider than with asulam alone. Weed grasses (see 5.14) will be checked.

5.23. **Mixtures of dicamba** + **mecoprop; dicamba** + **mecoprop** + **MCPA; dicamba** + **mecoprop** + **2,4,5-T** *for controlling docks in all-grass swards*

These mixtures should be applied when the docks are 15–20 cm high and growing actively. This may be at any time from April to mid-October but must be before flowering. If flowering has occurred, the docks should be defoliated and the regrowth sprayed – but wait for at least 14 days to allow for sufficient regrowth to absorb the spray. A second application is recommended when regeneration occurs.

Marsh horsetail. This species (*Equisetum palustre*) is a frequent cause of poisoning in horses; cattle and sheep are less affected. The usual source is contaminated hay. As it thrives on damp land, improving drainage will help to alleviate the problem.

Partial control can be achieved by using either MCPA, 2,4-D amine or 2,4-D ester (see 5.05), or some herbicide mixtures, e.g. 5.09. Spraying should take place when shoots have made maximum growth, but there is usually strong regrowth from underground parts in the year after treatment. Repeated spraying coupled with improved grassland management is the only way to long-term control.

Horsetails can be killed in grassland one week before ploughing up by spraying asulam at 2.2 kg ha^{-1} in late spring. Subsequent crops should not be sown for at least 6 weeks after treatment.

Common nettle. Common nettle (*Urtica dioica*), although rarely distributed across whole fields, frequently occurs as large clumps, particularly on high fertility sites near buildings or on fallow ground recently disturbed. Regular cutting will help suppress nettles but avoid close grazing in the spring. Unsightly patches can be treated by means of a knapsack sprayer, fitted preferably with a flood-jet to reduce drift.

Herbicides that can be used to kill nettles in grass/clover swards include: benazolin + 2,4-DB + MCPA (see 5.09); 2,4-DB ± MCPB (see 5.06); glyphosate (see 5.16).

Where clovers are not important, the following herbicides can be used: 2,4-D + MCPA (see 5.05); mecoprop (see 5.18); triclopyr (see 5.19).

5.24. **Mixture of 2,4,5-T + 2,4-D** *for killing nettles in all-grass swards*

Spray to run-off: thorough wetting of the foliage and stems is essential. Apply before flowering and repeat in the autumn as the nettles start to die down.

5.25. **Mixture of 2,4,5-T + 2,4-D + dicamba** *for controlling nettles in all-grass swards*

The best time to apply is at the flower-bud stage, *i.e.* just before the first flowers open, and when the weeds are growing actively. Later sprayings should follow topping of the pasture, as soon as there is sufficient regrowth to absorb the spray. Applying nitrogen one week before spraying will encourage grass recolonization. Sprayed nettles may become increasingly palatable to stock and may therefore be grazed preferentially. This could cause digestive upsets and grazing should be prevented until all top growth of the weed is desiccated.

5.26. **Mixture of mecoprop + asulam + MCPA** *for controlling nettles in all-grass swards*

Apply up to flower-bud stage.

Ragwort. Both common ragwort (*Senecio jacobaea*) and the less frequent marsh ragwort (*S. aquaticus*) are highly poisonous and regularly cause stock deaths. Ragwort is at its most dangerous condition in hay or silage. High priority should be given to the recognition and control of this weed (10). Common ragwort is specified as an injurious weed by order of the weeds Act 1959.

Ragworts develop into tall, erect plants producing conspicuous yellow daisy-like flowers. Common ragwort has finely divided leaves and large flat-topped inflorescences on thick stems. It occurs frequently on infertile, sandy or light-textured soils in both biennial and perennial forms. Marsh ragwort has less divided leaves and smaller, more regular inflorescences on shorter, slender stems. It is found on heavy, poorly drained soils and is more strictly biennial. Both kinds (referred to as ragwort hereafter) produce numerous parachute-like seeds which are dispersed over long distances, germinate readily in overgrazed swards and form seed heads in the spring of their second year.

Cattle, horses and to a lesser extent sheep, are all susceptible to ragwort poisoning. Early symptoms include low appetite, constipation, jaundice and straining. Animals do not normally eat ragwort when grass is plentiful; some may well develop a tolerance to the weed although freshly introduced stock may eat it avidly and suffer accordingly. However, the acceptability of ragwort is increased markedly by wilting, so cut stems should be collected and burned. Also, plants sprayed with herbicides are more likely to be eaten and animals should be kept out of sprayed fields until the weed is completely dead and, preferably, disintegrated or removed.

Pulling or digging up ragwort plants is not recommended except for light infestations that are uneconomic to deal with in any other way. Neither is the common practice of cutting flowering stems, since this may do more harm than good by encouraging prostrate, vegetative growth and regeneration from basal buds.

Heavy grazing by sheep for short periods in early spring will reduce a light infestation of ragwort. Even so re-invasion will occur if field management remains unchanged.

Unfortunately, the weed is resistant to MCPB and 2,4-DB and is only suppressed by clover-safe asulam at 1.1 kg ha^{-1} (see 5.14). Glyphosate may be applied selectively using a specified rope-wick machine (see 5.16). Where clovers are unimportant, MCPA, 2,4-D amine and 2,4-D ester (see 5.05) can be used effectively; if anything 2,4-D is slightly better than MCPA. Some herbicide mixtures also give adequate control including: asulam + mecoprop + MCPA (see 5.22) and dicamba + mecoprop + 2,4,5-T (see 5.23). All of these proprietary mixtures are devastating to clovers.

Timing of herbicide spraying is important. Delaying spraying until flower-bud formation gives good control of seedlings but indifferent control of second-year and older plants. Hence, for grazed swards, spraying earlier, when the older plants are still in the rosette stage, is now recommended (2). For fields intended for silage and hay, the best time to spray is in autumn, from mid-September to November, of the preceding year; this allows time for the weed to die and so reduces the risk of contamination.

Once herbicides have been used to reduce initial populations, repeat spraying, coupled with improved grassland management, will be needed as a preventive measure.

Rushes. Soft rush (*Juncus effusus*) is the most common species. Less widespread are hard rush (*J. inflexus*) and jointed rush (*J. articulatus*). Soft rush is distinguished from other rushes by having continuous pith inside its green flowering stems, which are devoid of green leaves (6).

Rushes are unpalatable to stock and are particularly persistent weeds, producing vast numbers of very small seeds which can be dormant in the soil for many years.

Rushes are ubiquitous in UK grassland, infesting approximately 30% of old swards on farms used for suckler cattle, a third of these cases comprising serious infestations (12). Poor drainage is commonly, but not solely, the decisive feature of rush sites. Infestations commonly arise where swards, weakened by disturbance or management and subject to flooding, give rush seedlings space to establish.

Control measures aimed at controlling/preventing rush infestation should include drainage as a prerequisite for most situations, coupled with the application of lime and fertilizer, and controlled defoliation to produce a vigorous sward; common rush is reduced considerably by regular annual mowing for hay.

Soft rush can be controlled by spraying MCPA, 2,4-D amine or 2,4-D ester (see 5.05) in late spring, just before flowering. Where clovers are important, selective control is only possible by applying glyphosate using a specified rope-wick machine (see 5.16). Very dense infestations containing many old and dead stems are best mown the previous season or early in the year, and the litter removed, so that fresh shoots of at least 4 weeks growth are sprayed. Mowing 6 weeks after spraying helps to improve control by encouraging a dense sward to develop.

Thistles. The main species growing in established grassland are creeping thistle (*Cirsium arvense*), spear thistle (*Cirsium vulgare*), dwarf thistle (*Cirsium acaule*), marsh thistle (*Cirsium palustre*), welted thistle (*Carduus acanthoides*) and musk thistle (*Carduus nutans*). For individual descriptions, see MAFF Advisory Leaflet No. 51, (8).

The most common thistle is creeping thistle, which persists and spreads chiefly by its underground, creeping roots. It can remain dormant for many years, producing new shoots when swards are weakened. Flowering stems appear in late spring, with flowers opening in July. Much of the thistle-down produced therefrom is devoid of seed but seeds can establish on open soil. Stems die back in late autumn.

Creeping and spear thistle are listed as noxious species. If allowed to grow unrestricted they are likely to spread to adjacent arable land. They restrict grassland productivity, either by depleting moisture and nutrient resources that would otherwise be used by grass, or by interfering with grazing or forage conservation. Because thistles are tall, conspicuous and rarely eaten by livestock, there is broad agreement amongst farmers that thistles are the most important weed of permanent grassland and that their presence should be minimized although, in the absence of authoritative data, judgements are often prejudiced by aesthetic considerations.

The widespread prevalence of thistles in grassland has been illustrated in several surveys, with infestations ranging from 20% in Northern Ireland to 50% in Scotland; in England and Wales 8% of permanent grassland was found to be heavily infested and a further 13% had incipient infestations (12). On the basis of these data, serious infestations occurred on 400,000 ha of grassland in England and Wales; including incipient infestations would raise the figure to over 1 million ha.

Thistles typify unbalanced grazing on soils of low fertility, involving overgrazing in winter and undergrazing in summer. Rotational grazing by cattle, or cutting for silage, combined with an increase in fertilizer nitrogen, can usually reduce an infestation, but frequent topping seldom eradicates the weed.

Several herbicides can be used on creeping and spear thistle, giving good control of top growth in the year of treatment and a useful long-term suppression, but no selective herbicide will eradicate these thistles with a single application.

Where clover retention is important, the following herbicides can be used: benazolin + 2,4-DB + MCPA (see 5.09); 2,4-DB or MCPB (see 5.06); glyphosate (see 5.16).

Spray any time when shoots are extending, up until most flower buds are well developed but not open.

For all-grass swards, any of the herbicides given in table 5.1, section (b) can be used, especially those containing MCPA and mecoprop. (Note also 3,6-dichloropicolinic acid, although this is expensive and therefore suitable mainly for spot-application.)

As swards age so the root system becomes more extensive and difficult to kill; one spray treatment is rarely sufficient. Time is needed for the herbicide to be translocated to the roots so allow at least 2 weeks before cutting or grazing.

Tussock grass/tufted hair-grass. This unpalatable grass (*Deschampsia caespitosa* grows in large clumps in low lying, poorly drained meadows. Improvement in drainage is the first step in its control.

5.27. **Dalapon sodium** *0.45 kg 145 l of water for spot-treatment of tussock grass in grass/clover swards*

Control is improved if the tussocks are defoliated closely about 6 weeks before spraying. The tussocks need to be thoroughly wetted with the herbicide.

References

1. Anon. (1979a). The control of ragwort. *Scottish Agricultural Colleges Publication No. 38*, 7 pp.
2. Anon. (1979b). Bracken control in grassland. *Scottish Agricultural Colleges Publication No. 39*, 6 pp.

3. Cooper, F. B. (1982). Experiences in controlling *Bromus mollis* in permanent swards. *Proceedings 1982 British Crop Protection Conference – Weeds*, **1**, 381–385.
4. Doyle, C. J.; Oswald, A. K.; Haggar, R. J.; Kirkham, F. W. (1984). A mathematical modelling approach to the study of the economics of controlling *Rumex obtusifolius* in grassland. *Weed Research*, **24**, 183–193.
5. Gutsell, R. J. (1982). Field experience in combining MCPA and fertilizers for the control of *Ranunculus* spp. (buttercup) in permanent pasture. *Proceedings 6th British Weed Control Conference*, 105–122.
6. MAFF. (1976). Weed control – rushes. *Ministry of Agriculture, Fisheries and Food (Publications). Advisory Leaflet 433*, 5 pp.
7. MAFF. (1977a). Weed control – docks. *Ministry of Agriculture, Fisheries and Food (Publications). Advisory Leaflet 46*, 6 pp.
8. MAFF. (1977b). Weed control – thistles. *Ministry of Agriculture, Fisheries and Food (Publications). Advisory Leaflet 51*, 6 pp.
9. MAFF (1978). Bracken and its control. *Ministry of Agriculture, Fisheries and Food (Publications). Advisory Leaflet 190*, 8 pp.
10. MAFF. (1982). Ragwort. *Ministry of Agriculture, Fisheries and Food (Publications). Advisory Leaflet 280*, 6 pp.
11. MAFF. (1983). Weed control in grassland, herbage legumes and grass seed crops 1983–84. *Ministry of Agriculture, Fisheries and Food (Publications). Booklet 2056 (83)*, 38 pp.
12. Peel, S.; Hopkins, A. (1980). The incidence of weeds in grassland. *Proceedings 1980 British Crop Protection Conference – Weeds*, **3**, 877–890.
13. Popey, A. I.; Hartley, M. J. (1979). Barley grass, control and evasion. *MAFF Research Division, Farm Production and Practice*, 3 pp.
14. Roberts, H. A. (1982). *Weed Control Handbook: Principles, Chapter 12, Weed Control in Grassland*, pp. 351–368. Oxford: Blackwell Scientific Publications.

Chapter 6

PESTS OF ESTABLISHED GRASSES
          AND LEGUMES

Several pests cause damage to established grassland. Usually the damage is insidious and is easily overlooked. However, insidious damage by frit fly, slugs and leatherjackets frequently results in total annual herbage yield losses of over 20% (3). The populations of several pests including chafers and army worms occasionally reach very high numbers locally. Obvious, sometimes even spectacular, damage can result.

Frit fly. (Plate Vg and h – see also Chapter 3.) The frit fly larvae (e.g. *Oscinella frit*) are most prevalent in ryegrass, especially Italian ryegrass (4). Some grasses, e.g. cocksfoot, are seldom if ever severely attacked. The damage threshold of the larval population for established swards is thought to be about 300 larvae per square metre, and in ryegrass is seldom below this figure. In Italian ryegrass, populations resulting in 30% of tillers being infected occur commonly. Yield losses of between 10 and 30% of total annual dry-matter production, largely resulting from damage by frit larvae, were found to occur in eight out of ten established ryegrass swards studied. The characteristic lack of persistence in Italian ryegrass is also at least partly attributable to damage caused by frit (1).
Satisfactory control measures for frit in established grassland have not yet been developed.

Leatherjackets. (Plate Vc – see also Chapter 3.) The larvae of leatherjackets (*Tipula* spp.) graze on roots and may cause bare patches. Damp, cool conditions in spring, which favour leatherjackets more than grass growth, exacerbate damage.
Chemical control probably becomes worthwhile when populations exceed 85 m^{-2}. Chlorpyrifos, triazophos and gamma-HCH are all approved for this purpose. DDT must not be used.

Slugs. (Plate Va – see also Chapter 3.) Slugs (e.g. *Deroceras reticulatum*) can be very numerous in grassland, especially in wet areas on heavy soils. No estimates of threshold populations above which economic damage occurs seem to have been published. However, control with either metaldehyde or methiocarb is approved.

Wireworms. (Plate Vd – see also Chapter 3.) Wireworms (*Agriotes* spp.) can be very numerous, especially in old pastures. It is, however, generally thought that they cause little damage to established grass and control measures are therefore thought to be unnecessary.

Chafer grubs. (Plate Vb.) Two species of Scarabaeid beetles (*Phyllopertha horticola, Melolontha melolontha*) are of importance in grassland. The most important species, the garden chafer (*P. horticola*), has a one-year life cycle. Eggs are laid mostly during June and the larvae which hatch after 4–6 weeks then feed on grass roots until October when they become quiescent, passing the winter in diapause. The larvae are typically curled or 'C' shaped, have long legs, a distinct orange or brown head capsule but are otherwise pale or white in colour. Often the dark gut can be seen through the semi-translucent body. Pupation occurs in April/May and the adults emerge in May and June. The adult beetles fly in warm, sunny weather, often forming a swarm to mate.

The cockchafer (*M. melolontha*) has a three-year life cycle. Eggs are laid in the soil, mostly in grassland, during late May and June. The eggs hatch after 3–6 weeks and the larvae, which resemble those of the garden chafer, feed on a variety of plant roots but have a preference for grass rather than broad-leaved weeds. The larvae take some three years to mature. Although they cause little damage when small, they may reach 40 mm in length in later years and can then be very destructive. The larvae pupate at the end of their third season, emerging as adults in April/May of the following year.

Both chafers are widely distributed, but damage by them is most likely to occur on light land, especially that over chalk or limestone and in hill farming regions. Attacks are often heaviest on sloping terrain or near the brow of a hill and larvae tend not to congregate where conditions are excessively wet or very dry. Cocksfoot is less liable to damage than other grasses.

Fine-leaved grasses are the most severely attacked. Grass near woodland may be more prone to attack. Damage can be severe enough to be easily noticeable and sometimes even spectacular.

Since the larvae are easy to damage physically by cultivations, control can be achieved by ploughing and re-seeding; this solution is not possible however in hill areas, where soil treatment with gamma-HCH may be necessary.

Antler moth or army worm. The caterpillars of these (*Cerapteryx graminis*) are found in hill pastures, particularly in the Pennines, Scotland and Wales. The caterpillars are bronze-brown and glossy with a wrinkled yellowish-white line down the sides and back. The moth larvae attack the roots and aerial parts of most grass species but feed chiefly on bent wire grass (*Nardus stricta*) (5). The adult moth is a dull greyish-brown. White markings on the forewings resembling a stag's antlers give rise to the common name of Antler moth. Eggs are laid in August and are simply dropped on to herbage by the female moth while in flight, each laying up to 200 eggs. Some eggs remain dormant until the following spring, but others hatch and pass through the winter as diapausing larvae, which feed from spring until June when they pupate, in the soil, for about 25 days.

Visible damage is caused only rarely by these larvae. However, in some seasons they may become locally abundant and outbreaks of the caterpillars can cause severe damage over areas of several hectares. The larvae sometimes migrate in large groups (hence the name 'army worm'), denuding hill land of plants as they move.

The mass migrations of larvae can be controlled by digging ditches with vertical sides in their path, by burning the grass when dry, or by using poison baits.

Aphids. (Plate Ve and f.) Several species of aphid including *Rhopalosiphum padi*, *Sitobion avenae* and *Metopolophium festucae* occur commonly in grassland. Some, e.g. *Acyrthosiphon pisum* feed on legumes. They are green or brown in colour, have pear-shaped or oval bodies, and are small, most species seldom exceeding 4 mm. Their mouthparts are adapted to pierce plants and feed on sap. They may be winged or wingless and pass through many generations in a season. Aphids often give birth to living young and their numbers can build up very quickly during favourable conditions in long dry periods of warm weather during May–August. Even so, except in herbage seed crops, they rarely cause significant damage because the normal defoliations associated with grassland husbandry remove the aphids along with the herbage. However, if defoliation is infrequent and conditions favour their rapid multiplication, aphids may cause patches of stunted growth.

Some aphid species can transmit viruses, particularly Barley Yellow Dwarf Virus, which can infect many grass species (see p. 69). The cocksfoot aphid (*Hyalopteroides humilis*), for example, transmits cocksfoot streak virus.

Since aphids cause little damage to grasses or legumes it is unlikely to be economic to spray for this reason alone, but if virus diseases are a problem it may be desirable. Dimethoate, demeton-S-methyl, oxydemeton methyl, pirimicarb or thiometon can be used to check their numbers. Defoliation by grazing, or better still by taking a silage cut, is however an effective and cheaper control measure.

Bibionids. The larvae of Bibionids (St. Mark's Fly – *Bibio marci* and Fever Fly – *Dilophus febrilis*) resemble small leatherjackets up to 1.5 cm long but have a distinct brown or black head. The larvae are legless and usually dull green or brownish in colour. They are common and often very numerous, but their distribution within a pasture is very aggregated, large numbers being found scattered through the sward in small patches of a few square centimetres. Adults emerge in April often on or about St. Mark's day, 25 April, hence the common name of the fly. Eggs are laid and the larvae feed during the following twelve months, pupating in the soil in March. It is not clear whether the larvae feed mainly on live roots or on dead plant material, but they can cause damage by loosening soil around the base of plants. Control is not thought to be necessary.

Clover weevils. Most legumes but not grasses are attacked by these weevils (e.g. *Sitona lineatus*) (2). The adults make the commonly seen U-shaped notches in the edges of clover leaves. The adults are typically weevil shaped, having a long snout. They are darkly coloured with faint stripes, are small (5 mm long) and although common are seldom seen. The adults hibernate, often in plant debris, and emerge in late spring. Females may lay hundreds of eggs, although many fail to hatch. The larvae live in and feed on root nodules and subsequently pupate in cells in the soil about 50 mm deep. Adults emerge from the pupae from July onwards but do not breed until the following year. Generally, forage legume crops easily withstand feeding by both adults and larvae. Although, rarely, severe damage can occur if large numbers of adults attack small legume seedlings, control measures are not usually thought to be necessary.

Grass moths. Grass moths (*Crambus* spp.) are common throughout the UK. The adult moths, which are greyish-white with a 20–30 mm wingspan, spend much of their time at rest on grass plants, and if disturbed fly only a short distance before settling again. A distinguishing feature is an obvious projection from the front of the head. There is probably only one generation per year in the UK, but there may be more, as in N. America (2). Eggs are laid in late summer, either dropped on to the soil surface or laid at the base of grass plants. The caterpillars hatch and feed mostly on underground plant parts but also on leaves, until the onset of cold weather. The larvae are whitish, often with a row of black dots along each side. They have a distinct dark coloured head capsule and short legs. Often long bristles are clearly visible protruding from the body. The larvae are sometimes known as web-worms because they live in burrows lined with silken webs; they remain quiescent during winter but feed again with the return of warmer weather, pupating in the soil in June. Adults emerge some weeks later.

Although common in lowland grassland as well as in hill areas, it is in the uplands that damage is usually observed. Outbreaks of grass moth larvae, which occur infrequently, can cause severe damage over areas of several hectares on hill farms. Attempting to control the outbreak once damage is obvious is usually futile, but damage can be averted by sowing more desirable herbage species, e.g. perennial ryegrass, which are less susceptible to attack.

Swift Moth. Caterpillars of the ghost swift moth (*Hepialus humuli*) are common in grassland. They have short legs, a distinct cutinised head and are white in colour. Adult moths are fairly large with a 40–65 mm wing span. Males are white, females a drab orange-yellow colour with a thick body. Both fly quickly, hence their common name. Between 200 and 700 eggs are dropped on to vegetation by each female while in flight during June-August. The eggs hatch in 12–18 days and the caterpillars, although very active, develop slowly, feeding on plant roots, usually taking two years to reach maturity. Occasionally they take a third year to mature, or if conditions are particularly favourable they may complete their life cycle in one year. The caterpillars often continue to feed during winter. The fully grown caterpillars, which are about 65 mm long, pupate in the soil in spring emerging as adults from the end of May onwards.

Control can be effected by thorough cultivation, which kills many of the larvae directly and enables parasitic fungi to attack others more easily .

References

1. Clements, R. O.; Henderson, I. F. (1983). Improvement of yield and persistence of Italian ryegrass through pest control. *Proceedings XIV International Grassland Congress, Lexington, June 1981*, pp. 581–584.
2. Edwards, C. A.; Heath, G. W. (1964). *The principles of agricultural entomology*. London: Chapman & Hall, 418 pp.
3. Henderson, I. F.; Clements, R. O. (1977). Grass growth in different parts of England in relation to invertebrate numbers and pesticide treatment. *Journal of the British Grassland Society*, **32**, 89–98.
4. Henderson, I. F.; CLements, R. O. (1979). Differential susceptibility to pest damage in agricultural grasses. *Journal of Agricultural Science, Cambridge*, **93**, 465–472.
5. Jones, F. G. W.; Jones, M. (1964). *Pests of field crops*. London: Edward Arnold, 406 pp.

Chapter 7   # DISEASES OF ESTABLISHED SWARDS

Part A   **Diseases of Grasses**

Diseases Caused by Fungi

Foot and root diseases

Take-all. Take-all (*Gaeumannomyces graminis*) is rarely a problem in its own right on agricultural grasses, except under continuous Italian ryegrass monoculture with heavy nitrogen dressings, carried over to a subsequent cereal crop. Infected grasses have dark runner hyphae together with slender infection hyphae on the root surfaces, and a black stromatal plate, less pronounced than in cereals, between leaf sheath and culm and bearing dark perithecia. No chemical control measures are available.

Fusarium root rot. The commonest species of *Fusarium* isolated from grass roots is *F. culmorum*. Perennial ryegrass seems worse affected than Italian, and a severe loss of stand can occur even when the sward is well beyond the establishment phase. Poor management, particularly inadequate drainage, is largely responsible. No chemical control measures can be recommended, although chlorothalonil, having proved effective against *Fusarium* in sports turf, might be of use here as well as against the foot rot phase of infection by *Drechslera* spp. (q.v.).

Ligniera junci. No common name exists for the disease caused by this root parasite, first described by Michail and Carr (16) as a possible cause of the premature disappearance of Italian ryegrass from young leys. Perennial ryegrass was little affected and so became dominant. Infected plants appear sickly, with reduced vigour and tillering, ultimately turning brown and dying. The root cortex contains abundant, brown resting sporangia (cystosori). This disease occurs most often in exceptionally wet conditions. Nothing is known regarding possible control measures.

Snow mould. This fungus (*Micronectriella nivalis*) is known more commonly by the name of its 'imperfect' stage, *Fusarium nivale*. Two distinct diseases are caused in grasses: fusarium patch, a turf disease found throughout Britain, and snow mould (often associated with *Typhula incarnata*) of agricultural grasses in upland Scotland. It has been estimated (13) that 150,000 ha of grassland in northern Scotland are continually at risk from snow mould, rising to 250,000 ha in severe winters. The disease affects bents, annual meadow grass and crested dogstail, but not red fescue or smooth stalked meadow grass. Among sown species, Italian and perennial ryegrasses are particularly at risk; cocksfoot and timothy are moderately so.

Snow mould causes maximum damage during periods of prolonged snow cover, particularly where drifting snow accumulates over unfrozen ground especially when previously treated with slurry or managed intensively. Isolated plants are observed having bleached, watersoaked leaves. Collapsed leaves usually have a weft of pinkish mycelium. Within days many tillers and even whole plants die, causing circular, dead areas and subsequent weed encroachment.

Improved drainage, sparing use of nitrogen and removal of surplus growth well before the onset of winter can all limit the development of snow mould. There is experimental evidence that benomyl (1 kg a.i.ha^{-1}) controls it (13), and quintozene has been reported effective. Cultivars vary considerably in tolerance to cold and resistance to snow mould. Dutch tetraploid ryegrasses have been found better than British diploids, though probably more because of origin than of ploidy as such. The best prospects lie in the active breeding programmes currently being pursued at several centres.

Stem and foliage diseases

Of the 20 or so rusts attacking grasses in Britain, only three are considered of sufficient economic importance to merit consideration here. All three have cereal hosts, but the available evidence indicates that the forms attacking grasses are incapable of infecting cereals.

Crown rust. This is undoubtedly the most important rust (*Puccinia coronata*) (and indeed one of the most important pathogenic fungi) infecting grasses in Britain (Plate VIa). Separate forms, f.sp. *lolii* and f.sp. *festucae*, attack ryegrasses and fescues; other forms attack agriculturally unimportant grasses such as tall oat, blackgrass and fog. The disease is recognized easily by the scattered, bright orange spore pustules (uredosori) on both leaf surfaces, from which the short cycle of infection originates. From September onwards black, linear teleutosori appear mainly on the lower surfaces, interspersed among the existing uredosori. Basidiospores produced from germinating teleutospores can infect only the alternate hosts, buckthorns (*Rhamnus* spp.); aecidiospores produced on buckthorns in spring can infect only the grass hosts, thus completing the long life cycle and ensuring genetic recombination and the possibility of new virulences.

The development of crown rust is favoured by a cycle of warm dry days and dewy nights, conducive to spore dispersal and subsequent infection; hence heavy attacks are commonest in late summer/early autumn and in some springs. Such attacks can reduce yield by 35%, rendering the grass unpalatable and less nutritious to stock, soluble carbohydrate being reduced by as much as 22%. Because of reduced height (by 10%) and tillering (by 20%), spring regrowth can be reduced by as much as 23% and competitive ability is also markedly affected, leading to clover dominance in ryegrass/clover swards (14) or other forms of sward imbalance.

Leys receiving more than 250 kg ha^{-1} of nitrogen are generally free from severe infection. Grazing or cutting more frequently than usual, particularly in autumn, helps to prevent disease build up. There are no firm recommendations for chemical control, but this has been achieved in experimental plots with nickel sulphate plus maneb, and with the systemic fungicides, oxycarboxin (200–400 g a.i.ha^{-1}), benodanil (1.1 kg a.i.ha^{-1}) and triadimefon (100–250 g a.i.ha^{-1}). Chlorothalonil has also proved partially effective, though its main action was in controlling drechslera leaf spots. Careful choice of cultivar together with sensible management seems to offer the best hope of control, particularly for the southern half of Britain where the disease is more prevalent and severe. Italian are generally more susceptible than perennial ryegrasses, earlier than later flowering cultivars, and diploids than tetraploids. Some have been bred specifically for crown rust resistance, notably some of the newer tetraploid hybrid ryegrasses.

Brown rust. Brown rust (*Puccinia recondita* f.sp. *lolii*) normally occurs less frequently than crown rust on ryegrasses, but has been found in epidemic proportions on breeding material introduced from abroad, notably the Po valley region of Italy (22).

LEGENDS

Plate I: Grasses.
  a. *Poa annua*, ligule
  b. *P. annua*, habit
  c. *P. trivialis*, ligule
  d. *Avena fatua*, inflorescence
  e. *Elymus (Agropyron) repens*, ligule
  f. *E. repens*, inflorescences
  g. *Hordeum murinum*, ligule
  h. *H. murinum*, inflorescences

Plate II: Grasses
  a. *Bromus sterilis*, ligule
  b. *B. sterilis*, inflorescence
  c. *B. hordeaceus (mollis)*, ligule
  d. *B. hordeaceus*, spikelets
  e. *Holcus lanatus*, inflorescences
  f. *H. lanatus*, ligule
  g. *H. mollis*, ligule
  h. *H. mollis*, habit

Plate III: Grasses
  a. *Arrhenatherum elatius* ssp. *bulbosum*, ligule
  b. *A. elatius* ssp. *bulbosum*, spikelets
  c. *A. elatius* ssp. *bulbosum*, culm bases with bulbous short internodes
  d. *Anthoxanthum puelii*, ligule
  e. *Phleum pratense*, inflorescences
  f. *P. pratense*, ligule
  g. *Agrostis capillaris (tenuis)*, ligule
  h. *A. capillaris*, habit

Plate IV: Grasses
  a. *Alopecurus myosuroides*, ligule
  b. *Agrostis stolonifera*, ligule
  c. *A. stolonifera*, habit
  d. *A. gigantea*, ligule
  e. *Avena fatua*, ligule
  f. *Agrostis gigantea*, habit

Plate V: Pests
  a. Slug, *Deroceres* sp.
  b. Chafer grub, *Hoplia philanthus*
  c. Leather jacket, *Tipula paludosa*
  d. Wire worm, *Agriotes* sp.
  e. Grain aphid, *Sitobium avenae*
  f. Aphid on fescue leaf, *Rhopalosiphum* sp.
  g. Frit fly, adult, *Oscinella* sp.
  h. Frit fly, larva, *oscinella* sp.

Plate VI: Diseases of Grasses
  a. Crown rust of ryegrass, *Puccinia coronata* f.sp. *lolii*
  b. Net blotch on rye grass, *Drechslera dictyoides*
  c. Choke on cocksfoot, *Epichlöe typhina*
  d. Bacterial wilt of ryegrass, *Xanthomonas campestris* pv. *graminis*
  e. Ryegrass mosaic virus (RMV)
  f. Barley yellow dwarf virus (BYDV) in an Italian ryegrass plant (R); healthy plant (L)

Plate VII: Diseases of Clovers
  a. Clover rot causing a stolon rot in white clover, *Sclerotinia trifoliorum*
  b. Clover scorch in red clover, *Kabatiella caulivora*
  c. Downy mildew on white clover, *Peronospora trifoliorum*
  d. Red clover necrotic mosaic virus (RCNMV) causing crinkling and necrosis of the foliage of red clover
  e. Clover phyllody causing leafy proliferation of the inflorescences of white clover. Healthy inflorescence on right
  f. English stolbur (Red leaf disease) on white clover

Acknowledgements

Plates I–IV. Reproduced, with permission, from the BASF publication *Grass Weeds in World Agriculture* by S. Behrendt and M. Hamf
Plate V g. © Ivor J. Dixon/LP and TS
Plate V a–e. By courtesy I.C.I. Plant Protection Division
Plate V f and h. By courtesy Grassland Research Institute
Plates VI and VII. By courtesy Welsh Plant Breeding Station

Plate I

a

b

c

d

e

f

g

h

Plate II

Plate III

Plate IV

Plate V

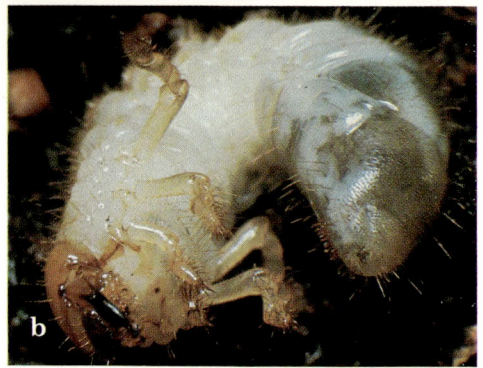

Plate VI

Plate VII

Unfortunately genetic resistance to *P. coronata* does not confer resistance to *P. recondita*. The scattered pustules occurring mainly on the upper leaf surfaces are distinctly browner in colour than those of crown rust. Most of the recommendations given for control of the latter apply also here.

Stripe rust. Stripe rust (*Puccinia striiformis*) is the most serious rust infecting cocksfoot, especially in warm dry years and on mature growth, from mid-summer on into autumn. Heavy infection renders the grass unpalatable and markedly reduces soluble carbohydrate. Pustules, both the lemon-yellow uredosori and the later formed, dark teleutosori, occur in characteristic long lines which give the disease its name, on leaves, sheaths, culms and inflorescences. Cultivars vary broadly in their susceptibility. Control by protectant (e.g. nickel sulphate plus maneb) and systemic (e.g. benodanil) fungicides has been claimed, but there are no recommendations. Again, reasonably frequent cutting or grazing combined with a good level of nitrogen will help keep the disease in check.

Powdery mildew. Attacks of powdery mildew (*Erysiphe graminis*) are common on most grasses in Britain as well as on cereals, though because of marked specialization forms on grasses play no significant role in infecting cereals, or *vice versa*. The disease is favoured by warm, dryish periods in spring and summer, which may explain why it seems more common in the eastern half of Britain than in the west, particularly on cocksfoot but also on ryegrasses, especially Italian. Mildew is worst under conditions of high soil nitrogen, and is most likely to damage short-term leys and mature growth, the dense canopy affording the right conditions of shade, reduced air circulation and a resultant high humidity. At high levels of ryegrass infection, where mildew is evident on the third leaf, losses of 20% have been recorded and quality is reduced by increase in dead material.

Whitish, powdery pustules appear on the upper leaf surfaces from early spring onwards, the entirely superficial mycelium bearing long conidial (spore) chains. The pustules later turn a buff brown colour as the leaf itself turns yellow. Sporulation is so profuse and the disease cycle so rapid (a few days only) that mildew builds up to an epidemic faster than any other fungal disease of herbage. Later in the growing season small dark spherical structures (cleistothecia), containing the sexually produced ascospores, appear within the mycelial weft. These, and viable mycelium on living host tissue, assist in overwintering and maintain genetic variation.

Reducing the canopy density by grazing or cutting allows light to penetrate and air to circulate, and assists in controlling mildew. There are no chemical control recommendations, though in field trials the systemic fungicides tridemorph, triadimefon and benomyl have proved extremely effective. Tridemorph has been known to have some phytotoxic effect, and is possibly best used at about half the application rate recommended for use on cereals. Some resistant Italian ryegrass cultivars are available, both diploid and tetraploid. The latest NIAB ratings should be consulted (see p. 5).

Rhynchosporium leaf blotch. Leaf blotch (*Rhynchosporium secalis* and *R. orthosporum*) provides one of the few examples of a fungal disease where there may be an epidemiological connection between grass and cereal hosts (21). Both species infect Italian ryegrass, *R. orthosporum* infects cocksfoot and timothy.

The disease is most noticeable in a cool, moist spring, particularly in the west and south-west. Irregularly shaped blotches up to 25 mm in length appear on the leaf blades and sheaths near the auricles ('spring burn'). Water soaked at first, they later turn grey and develop a dark brown margin, the grey centres becoming covered with spores which are disseminated by rain splash. Leaf blotch, in combination with mildew, can reduce yield of a susceptible Italian ryegrass cultivar by 25% (10). Some of the more popular cultivars are very susceptible. Fortunately, others have more resistance, notably the tetraploid Italian and the new tetraploid hybrid ryegrasses. Use of these, and taking an early cut, should afford adequate protection, particularly as leaf blotch, being a spring disease, is less likely to appear in the regrowth. Chemical treatment is probably not necessary, and none is recommended, although some success has been reported with benomyl.

Drechslera leaf spots. Three *Drechslera* species attack ryegrasses and fescues (23): *D. siccans, D. catenaria* and *D. dictyoides* (Plate VIb). *D. siccans* is commonest on ryegrass, forming dark brown oval lesions on the leaves, but not to the exclusion of the other two; this species also infects fescues and cocksfoot. *D. catenaria*, like *D. dictyoides*, produces a brown reticulation on the leaves (net blotch) rather than discrete lesions, and occurs on cocksfoot in particular but also on fescues and ryegrasses. *D. dictyoides* is believed to exist in two forms, a non-protothecial producing form, f.sp. *perenne*, attacking ryegrass, and a protothecial form, f.sp. *dicytoides*, attacking fescues. However Lam (personal communication) considers f.sp. *perenne* to be a separate species (*D. andersenii*) and has isolated two other species, *D. nobleae* and *D. sorokiniana* from ryegrasses. Another species, *D. festucae*, infects tall fescue, giving rise to chocolate brown lesions up to 30 mm in length (20). *D. phlei* causes an extensive leaf streak of timothy and also attacks cocksfoot. Most of these *Drechslera* spp. are seed borne (18). *D. siccans* can, under wet conditions, kill whole tillers and a serious footrot may develop.

Drechslera spp. cause considerable damage, on ryegrasses ranking with crown rust and rhynchosporium leaf blotch as the most important fungal diseases. They are particularly severe from late summer through the winter and into early spring, the cool wet conditions favouring spore dispersal and infection. Late in the season the leaves turn yellow and brown at the tips ('winter burn'). Yield losses of 10% with only 5% lesion cover have been recorded. Nitrogen fertilization appears to increase disease incidence. Sward composition may be affected, other species taking the place of the affected ryegrass.

There is little information on disease control, other than to cut or graze infected swards as appropriate to limit disease build up. A 12% increase in yield over infected swards can result from application of chlorothalonil one month after sowing and repeated at 4–6 week intervals throughout the growing season. Tetraploid ryegrasses, both Italian and perennial, are the most resistant (23), though as yet there are no positive NIAB recommendations.

Cocksfoot leaf fleck. Cocksfoot leaf fleck (*Mastigosporium rubricosum*) is the commonest and also one of the most destructive diseases of cocksfoot, particularly in the wetter west, reaching a peak during the cool, moist weather of early spring and again in autumn. Mean leaf infection levels commonly reach 5%, and on the lower leaves 25% or more is by no means uncommon, when premature leaf shed often occurs. Carr (4) refers to infection levels of 10% leaf area cover causing a loss of 50% of available soluble carbohydrate. Infection appears first as elliptical, water-soaked spots, about 1 mm long, on both sides of the leaf. These enlarge to 3–6 mm long purplish brown flecks, often surrounded by a tan margin. The centres tend to bleach and in damp weather show pinpoint white dots consisting of glistening masses of rain-dispersed conidia. *M. rubricosum* also infects timothy and perennial ryegrass; timothy is infected also by *M. cylindricum,* and *M. album* can infect perennial ryegrass.

Mature herbage suffers more severely than young growth, so cutting or grazing frequently curtails disease development. A late cut reduces overwintering. Adequate potash levels tend to lessen disease incidence but heavy nitrogen applications will increase it. There are no recommendations for chemical control.

Halo spot. Halo spot (*Selenophoma donacis*) infects a number of grasses, principally timothy and cocksfoot, and is seed borne on many of them (18). Oval, purple bordered spots appear on the leaf blades and sheaths of timothy in late autumn, often giving the leaves a complete purple appearance. The lesions enlarge and often run together to form irregular, longitudinal streaks up to 50 mm. The lesion centres bleach to a pale straw colour and bear rows of small, dark brown round bodies just visible to the naked eye. These are the pycnidia, from which, under moist conditions, pycnidiospores ooze in a slimy thread to be dispersed by rain, so that the damage is worst in a cool wet season.

Symptoms on cocksfoot usually appear later in the year, when the plants are well into head. The lesions, which are smaller and more diffuse than those on timothy, lacking a purple border, develop not only on the leaves but also on the rachis and often on the panicle branches and glumes, affecting the yield and quality of seed. Isolates from either grass host will infect the other, though each still tends to show a preference for the host species from which it was isolated. Those from barley will not, however, infect grasses, nor will those from grasses infect barley, so there is no risk of carry over of infection to the cereal crop. High nitrogen increases disease levels. There is rarely a spring or autumn when the disease is not present, at least on more mature growth, though it tends to disappear temporarily during mid season.

There are no control recommendations other than the obvious one of organizing the cutting or grazing regime to limit the build up and carry over of infection.

Cladosporium (Heterosporium) leaf spot of timothy. This disease seems to have increased over the years to a level at least equal to halo spot, with which it can easily be confused. Again, mature symptoms consist of scattered, oval and sometimes coalescing lesions, each about 3 mm long, having fawn centres and intensely purple margins. However, the dark spots in the centre of each lesion are not pycnidia, but dark conidiophores bearing the characteristic dark spores of *Cladosporium*. The disease tends to occur rather later than halo spot, towards mid season, and can do considerable damage. Japanese workers have estimated losses in crude leaf protein due to *C. phlei* of 25%. No control measure is known.

Smuts. Smuts are ubiquitous pathogens, infecting, according to species, leaves, stems or inflorescences. They are rather rare and therefore of little importance on modern agricultural grass cultivars, but because of the symptoms even just the occasional smutted plant is strikingly obvious.

Stripe smut (*Urocystis striiformis*) affects several grasses, including ryegrass, cocksfoot, timothy and red fescue. It is systemic throughout the plant and both soil and seed borne. The leaves of infected plants bear long, yellow-green streaks which turn grey; the cuticle and epidermal cells covering them rupture, exposing the underlying black spore masses, while the leaves split into ribbons, curl down, brown and die.

Flag smut (*Urocystis agropyri*) produces similar symptoms, commonly on couch and tall oat grasses but occasionally on timothy, ryegrass and meadow grasses.

Loose smut is a head smut very common on tall oat grass. It is caused by the same fungus (*Ustilago avenae*) responsible for loose smut of oats, but the one will not infect the other. The seeds are replaced by a mass of black compact sori yielding greenish brown teleutospores.

There is little information on the control of grass smuts. However, work in the USA where leaf smuts are a greater problem than in Britain has shown that flag smut can be eradicated by soil application of benomyl, oxycarboxin, thiabendazole or 2,5-dimethyl-3-furylanide.

Inflorescence diseases

Blind seed disease of ryegrass (*Gloeotinia temulenta*). Perennial ryegrass suffers most, but infection of Italian ryegrass is not uncommon and the full host list includes cocksfoot, timothy, fescues, bent, meadow grasses and many wild species, particularly in wetter areas which favour the rapid spread and recycling of the disease. In the Netherlands seed losses have been estimated at 20%, in epiphytotic years in New Zealand seed germination was reduced by 30%, and in Northern Ireland nearly 50% of inviable seed of some cultivars has been recorded.

The sequence of primary and secondary infection is identical; wind-borne ascospores, or rain-dispersed macroconidia, alight on the stigmas while these are exposed for pollination. They germinate and the resultant hyphae invade and often destroy the embryo, rendering the seeds 'blind' (26).

Control consists of sowing clean seed, or seed stored for at least two years as the fungus does not survive that long. Seed should not be taken from infected crops, although valuable seed can be treated with hot water, 50° for 30 min, or 45–46° for 2–2.5 h, to eradicate the fungus. Burning the straw and stubble will also assist provided there are no close sources of re-infection, as will drilling to a depth of 13 mm. Australian work has shown that benomyl applied to the soil at 2.9 and 5.6 kg ha^{-1} reduced apothecial production in perennial ryegrass seed stands by 80% and 90% respectively (15). An alternative approach is to breed, either for disease escape or for true resistance. Both approaches have been attempted with some success, but there are no truly resistant cultivars yet available.

Ergot. Most grasses as well as cereals such as rye and wheat are attacked by this most widely distributed and important of inflorescence diseases (*Claviceps purpurea*). Ryegrass is particularly badly affected.

The disease first appears on infected inflorescences about two weeks after ear emergence, as a sticky 'honeydew' ooze, containing the conidia which are spread by contact, insects and rain droplets to adjacent florets and those on other plants. The characteristic, black horned ergots (sclerotia) form in infected inflorescences from late summer to autumn, and fall to the ground where they remain dormant throughout the winter. They germinate in spring to produce flesh coloured fruiting bodies (stromata) in which the ascospore-containing structures (perithecia) are embedded. Spores of either type germinate only on the stigmas, generally of unfertilized florets, and infect the ovaries, giving rise once again to the 'honeydew' stage and later, to the ergots.

Ergot is important because of possible loss of seed, but more seriously because of the danger of disorders in cattle and sheep through ingestion of the alkaloid-containing ergots. These disorders are, mainly, gangrene of the extremities and convulsions of the nervous system leading to lameness; the claim of commonly occurring abortion due to ergot poisoning has probably been overstated. Minimum levels of ergot in feed required to cause clinical symptoms are set as low as 0.5–3% of total dry matter. These levels are achieved more readily by feeding silage rather than hay made from over mature pastures, or by direct grazing these; with hay more of the ergots are lost in the various harvesting and handling processes.

Control can be exercised by mowing old pastures well before heading, extending the time between susceptible crops to a minimum of three years, and before reseeding (with clean seed) deep ploughing so as to bury the ergots (at least 25 cm deep). Seed should not generally be taken from infected crops but contaminated seed can be cleaned by floating off the ergots in brine. Soil treatment with benomyl and related systemic fungicides has been found to control the disease both in infected grasses and by preventing development of the fruiting bodies from the shed ergots, but there are no recommendations for control by this method.

Choke. Choke (*Epichloe typhina*) is systemic in the vegetative parts in all of its 26 recorded grass hots, which include cocksfoot (Plate VIc), fescues, timothy, bents and meadow grasses but not ryegrass. The fungus breaks out on to the plant surface only in spring, converting the inflorescence into a fungal stroma in the form of a gradually thickening collar. White at first, and bearing colourless conidia, this turns yellow and then orange, containing immersed fruiting bodies (perithecia) from which ascospores are discharged forcibly. Conidia and ascospores can both germinate on cut plant surfaces, the mycelium passing down the pith and producing fresh stromata in the following season. Inflorescence emergence is prevented completely on most grasses, including cocksfoot, the stromatic sheath binding the leaves together.

Cocksfoot seed stands suffer the worst, showing an overall mean loss of seed due to choke of 3%, with 7% loss by the second harvest year and 21% in 5–7 year old crops. Timothy is less affected. There appears to be little effect on vegetative growth and even some increase in tillering with infection.

No chemical control measures can be recommended. However, gibberellic acid sprayed early in the season has been used successfully to obtain valuable breeder's seed from infected plants. For practical purposes, rogueing is partially successful. It is best not to grow cocksfoot for seed for more than 2–3 years, and old severely infected stands should be well ploughed in.

Diseases Caused by Bacteria

Bacterial wilt of ryegrss. (Plate VId.) This diseasse (*Xanthomonas campestris pv. graminis*) is apparently of recent origin, having been discovered in 1975 in Switzerland, Germany and France (12). It has since been found in Britain (24) and in Norway. Most reports have been of ryegrasses, particularly Italian and Italian/perennial ryegrass hybrids, being infected in breeders' and National List trials; very few have been of farm crops. In trial plots, levels of 20% infection have been recorded, and one of 80% in Italian ryegrass in its third harvest year. Cocksfoot, meadow fescue and timothy have also all been found as natural hosts while other species such as tall and red fescues have succumbed in inoculation tests.

Symptoms appear on ryegrass in early July and are most characteristic on flowering tillers. The leaves, especially the flag, have marginal yellow stripes which become progressively necrotic, the whole leaf curling, wilting and dying. Complete shoots may become stunted and die at about heading stage. Inflorescences have difficulty in emerging. Of those that do, a proportion become bleached, these 'whiteheads' looking very conspicuous. Affected leaves on vegetative tillers develop dark green, water-soaked spots which coalesce until the whole leaf becomes dark green before wilting, withering and dying.

Spread of the disease is facilitated by cutting, bacteria being easily transmitted to other plants via the cutting bar. It is probably also spread by grazing animals, and seed transmission is also suspected. Warm weather seems to foster symptom development, though not necessarily initial infection. High nitrogen increases incidence as does a prolonged period of water stress.

There are no specific control recommendations, though it would make sense to disinfect cutting machinery before moving on to the next crop when bacterial wilt has been diagnosed. Ryegrass cultivars vary markedly in their resistance: some tetraploids have shown quite high levels and could be exploited for disease control.

Diseases Caused by Viruses

Of the 22 viruses found infecting grasses in Britain (6,7) only five are sufficiently prevalent and serious in their effects on grasses of agricultural importance to merit consideration here. Some are sap-transmissible, that is, they can be transferred by rubbing sap from diseased on to healthy plants. This can occur when the crop is trampled or chewed, or is cut and crushed by farm machinery. All are also transmitted by specific organisms (vectors) which acquire them when feeding on infected plants and pass them to healthy ones. Some can be transmitted only in this way. Sap-transmissible viruses generally have a very transient existence in their vectors (non-persistent), whereas those that are transmitted only by vectors are generally retained in the vector for longer periods or even for life (persistent, circulative) and may even multiply there (persistent, propagative). Persistent viruses after being acquired by the vector go through a latent period before it can pass them on, non-persistent viruses do not. All these factors influence the ways in which viruses are spread and how control can best be exercised.

Some of these viruses have a very narrow host range, perhaps only a few grass species; others can infect a wide range of graminaceous plants, both grasses and cereals, and sometimes dicotyledons as well. With these, grasses often act as a perennial reservoir providing a constant source of virus infection to the annual cereal crops. All are systemic, that is, they are distributed throughout most of the host tissues. Once infected, the plant cannot be 'cured'. Accurate diagnosis requires considerable expertise, and access to facilities such as a range of vectors and diagnostic test plant species, electron microscopy and serology. Symptoms in themselves provide only a rough guide as quite different viruses can produce similar symptoms in the same host while, conversely, the same virus may produce dissimilar symptoms in different hosts. Sometimes symptoms are barely visible, yet the plants are infected and may suffer quite considerable yield losses.

Ryegrass mosaic virus (RMV). Ryegrass mosaic (Plate VIe) is generally regarded as the most serious of the viruses infecting grasses, if not the most serious of all grass pathogens, particularly in ryegrasses, Italian ryegrass being the most affected. Cocksfoot, meadow fescue and oats are rarely infected naturally. In ryegrasses, a faint, yellow-green mosaic develops on the leaves in late spring, some two weeks after infection, rapidly becoming more pronounced, and sometimes accompanied by a brown spotting, or even a more general browning (necrosis). Mild infection results in fewer tillers but plant height is not markedly affected and assuming a not too high infection level uninfected plants in the sward may make sufficient growth to compensate. Severe, necrotic infection not only reduces tillering but also leads to dwarfing and death. In this case, compensation by healthy plants is unlikely to be sufficient to avoid serious yield loss.

RMV is readily sap transmitted, so one would imagine that it could be spread easily by farm machinery or grazing animals. This is claimed not to be the case, most natural spread being reported to be by the only known vector, the Eriophyid mite, *Abacarus hystrix*. The mite inhabits the grooves between leaf veins and is not very mobile, but can be blown by the wind over considerable distances. It seems to dislike RMV infected plants, and so is dispersed more rapidly from infected than from healthy plants, assisting virus spread.

The disease is widespread and severe in the south, but infections have been seen as far north as the Orkneys. Most ryegrass seed stands in southern England have been shown to be infected; 13% of Italian ryegrass stands had from 30%–100% of plants infected and 10% of perennial ryegrass stands had over 20% of plants infected. Heavily infected swards can lose up to 30% of their herbage production; 15% is commonplace. Equally important, infected herbage is reduced in organic matter, percentage digestibility and soluble carbohydrate. The effects appear most marked under high nitrogen regimes, largely because infected plants fail to respond to the additional fertilizer (1). The virus also accentuates winter kill.

Control of RMV depends on restricting the spread of infection within the crop, destroying the mite vectors, and the use of resistant cultivars. Autumn sowing avoids the period of greatest mite dispersal (June–August). Workers in ADAS have found that Italian ryegrass sown in autumn had less than 5% of infected plants whereas sown in spring, under cereals, over 75% were infected. Corresponding seed yields were 308 and 196 kg ha^{-1}. Early and late grazing, especially by sheep, can help to reduce the mite population. Mites can also be controlled effectively by acaricides, e.g. aldicarb applied as a granular formulation to the soil, but this is unlikely to prove economic except possibly on seed production stands. Old, diseased swards should be killed completely before reseeding; the disease is not seed borne. Slot seeding for pasture renovation is likely to increase the risk where the old pasture is heavily infected.

Undoubtedly, the best prospect for control lies in the development and use of resistant cultivars (8). Tetraploid Italian ryegrasses have been developed with superior resistance compared with diploid cultivars, and some of the diploid late flowering perennial ryegrasses such as the WPBS cultivar, Mascot, are also resistant. NIAB ratings are at present given only for the Italian ryegrass cultivars. None of these is immune, but some have post-infection resistance. Plant breeders are currently combining resistance of this type with resistance present in genotypes derived from old pastures which renders them much more difficult to infect. This should confer even higher resistance levels to a whole range of new cultivars in prospect. There is also some prospect of breeding for escape from the mite vector: plants with leaves having shallow grooves have been shown experimentally to support smaller mite populations than those with deeper grooves which afford the creatures a better environment for development and survival.

Barley yellow dwarf virus (BYDV). (Plate VIf.) This is probably the commonest of all the grass-infecting viruses, although because the symptoms are not always obvious it is frequently overlooked. The distribution is world wide and the host range includes virtually all grasses and cereals. Grasses provide a perennial source of infection of annual cereals, the virus being transmitted within and between these crops by a large number of aphid species, the most important vectors being the bird-cherry aphid (*Rhopalosiphum padi*), the grain aphid (*Sitobion avenae*), the blackberry aphid (*S. fragariae*) and the rose-grain aphid (*Metopolophium dirhodum*). BYDV cannot be transmitted other than by aphids, in which once acquired it may persist for several weeks. At least half an hour is required for the aphid to acquire virus from an infected plant but it is unable to transmit it by feeding on another until a further 24–48 h have elapsed. Experiments at Aberystwyth have shown that during the autumn peak of aphid flight some 10–20% of the vector population caught in aerial traps are actually carrying BYDV.

Symptoms in ryegrasses and fescues are similar to those in cereals but are seen generally only on the flowering tillers, as a yellow discoloration of the foliage turning to red or purple. This begins at the leaf tip and extends downwards until most of the leaf blade is affected. Swards that are defoliated regularly rarely show these symptoms, and even in spring-sown hay and seed crops they do not usually appear until the second year. Timothy never produces anything other than a mild, inconspicuous chlorosis and infected cocksfoot is symptomless. The main effect of BYDV is to induce stunting and an increase in tillering, although the number of fertile tillers is decreased, markedly affecting sward dynamics (5). Greatest yield losses occur under conditions of grazing or frequent defoliation, which encourage the survival and competitiveness of the high tillering but poorly productive infected plants at the expense of the healthy ones. Infection also increases leaf soluble carbohydrate.

In plot trials, pure perennial ryegrass swards suffered a 25% yield reduction when half the plants were infected. In similar plots with white clover, the clover tended to become dominant. It is not certain what infection levels are reached in farm crops, but surveys have shown over 90% of ryegrass seed production stands to be infected in some years. BYDV in grasses, as in cereals, is probably more prevalent in the west and south than in the east of Britain.

It is difficult to foresee control measures which will overcome BYDV infection. Aphids, as pests, can be controlled as in cereals by appropriate insecticide sprays, dimethoate (335 g ha^{-1}), demeton-S-methyl (245 g ha^{-1}), oxydemeton-methyl (240 g ha^{-1}), pirimicarb (140 g ha^{-1}) or thiometon (275 g ha^{-1}). This treatment will not necessarily effect a total control of the virus however, because in order to be killed the aphids have to feed and on feeding will transmit the virus. Spraying will therefore not limit the introduction of infection, although it will control spread within the crop. Slot seeding to renovate old infected pastures will exacerbate the problem, as will completely reseeding the grass or sowing a cereal crop unless the old crop has been destroyed completely. *R. padi*, in particular, is a root feeder and transfers quite readily from plant debris to the new crop. Catherall and Wilkins have reviewed the problems of breeding ryegrasses for resistance to BYDV (9). Infected plants from early flowering cultivars actually outyielded comparable healthy ones in spring, but suffered much more in autumn. Further, decreases in height were partly compensated for by increased lateral growth so that the extent of yield loss is not necessarily a true indication of the level of susceptibility. This combination of changed seasonal response and agronomic type has thus far thwarted all attempts to select ryegrasses with durable resistance to this virus, although attempts to find a less environment-dependent form are continuing.

Ryegrass chlorotic streak virus (RCSV). Isolates of virus transmitted specifically by *R. padi* and no other aphid, once regarded as part of the barley yellow dwarf virus complex, are now designated ryegrass chlorotic streak, serologically distinct from BYDV although similar in many other properties. Generally, these *R. padi* transmitted strains are more severe, and although the symptoms in cereals are identical with those caused by BYDV those in ryegrasses are rather different, consisting of pronounced yellow streaks against a background of a dark blue-green leaf coloration. As with BYDV there is marked stunting and an increase in tillering but a reduction in the number of fertile tillers. Because of the similarity of the two viruses no separate assessment of the distribution and effects on yield can be given.

Cocksfoot streak virus (CSV). Cocksfoot streak, which infects only cocksfoot naturally but can be inoculated to ryegrasses, is a sap transmissible virus spread by the peach-potato aphid (*Myzus persicae*), and other aphids, in the non-persistent manner. It causes pale green or yellow streaks on the leaves of cocksfoot which can range from barely visible to quite distinct. It is widespread and common in seed crops, though perhaps rarer in leys. In many ways it is the antithesis of barley yellow dwarf, reducing the number of vegetative, but not the number of flowering, tillers without affecting plant height; hence when the sward is subject to frequent defoliation or grazing, infected plants are eliminated by competition from healthy ones (5). However, in hay and seed crops CSV infected plants have a competitive advantage over their healthy neighbours and a reduction in yield occurs. Other than managing the crop by the guidelines given above there are no specific recommendations for control. No resistant cultivars are currently available.

Cocksfoot mottle virus (CfMV). This virus is the most serious of those affecting cocksfoot in its effect on the crop, and is locally epidemic in Britain. Unlike the others reviewed here it is lethal, with the result that infected plants die out, to be replaced by other species and sometimes by weeds. Cocksfoot is the main natural host; occasional infection of wheat has been noted and both can easily be infected by inoculation with infected sap. The virus can be spread readily in the sward by farm machinery and grazing animals, and less successfully by adults and larvae of two beetles, the cereal leaf beetle (*Oulema melanopus*) and one which is commoner in grasses, *O. lichenis*. CfMV causes a marked yellow or white mottle on the leaves, intensifying into bleaching or necrosis of the older leaves, reduced tillering and plant height, and eventual death. Disinfection of cutting machinery would help to limit spread between crops, but could not be applied frequently enough to prevent spread within. No other control methods are available; however, the cocksfoot cultivar Cambria has a high level of resistance.

Part B   **Diseases of Herbage Legumes**

Diseases Caused by Fungi
Crown and root diseases

Clover rot. Clover rot (*Sclerotinia trifoliorum*) (3), although in its initiation perhaps not strictly a crown disease, normally reaches this region of the plant and causes maximum damage (Plate VIIa). Historically, it has been known as the most serious disease of red clover in Europe since the mid-nineteenth century, particularly in short rotations and on the more susceptible, double cut broad red types. Surveys in Britain in the late 1950s showed that 70% of red clover fields were infected, with losses of over 24% in 10% of crops.

Clover rot appears in autumn, leaves becoming peppered with small (less than 1 mm) brown spots as a result of ascospore infection. These may remain dormant for some months but in favourable weather, including frost, which lowers host resistance, these primary ascospore lesions enlarge and coalesce. The later stages of development, in which the leaves turn brown and wither, are favoured by milder moister weather occurring in late winter/early spring, the progress of the rot being often checked temporarily by severe cold. Ultimately, large, irregular patches of black, irregularly shaped resting bodies (sclerotia) are produced on and in the crowns and lower portions of the now well rotted plants. Sclerotia are released into the soil as the plant tissues decay, and most germinate not earlier than the following autumn, to produce one or more pinkish buff fruiting bodies (apothecia) having slender stalks with a disc- or funnel-shaped termination. Ascospores are discharged forcibly from these and are conveyed by the wind to infect the leaves, thus completing the disease cycle. There are no other viable spore forms.

Sclerotia can remain viable in soil for 5 years or even longer, only a proportion germinating each season. Susceptible crops such as red clover have been claimed to stimulate apothecial production from sclerotia. Additionally, primary sclerotia can produce secondary ones, particularly at soil moisture holding capacities in excess of 30%, although high soil moisture increases sclerotial decay.

S. trifoliorum also attacks other legumes, where until recently it was not regarded as particularly serious, and the common advice on badly contaminated land was to substitute the much more resistant white for red clover. However, during the past 15 years there have been serious outbreaks in white clover (2) and in some cases the damage has exceeded that on red clover.

Crop rotation is a general recommendation for control, 4–5 years between susceptible crops usually being adequate. Deep ploughing to bury the sclerotia below the maximum germination depth (50 mm) will also assist. Adequate seed cleaning and the use of clean seed is also important, as sclerotia carried with the seed could lead to fresh infection. However, tests have shown that the viability of sclerotia under 2 mm, the size which might remain after normal seed cleaning operations, is less than 5%. Clover rot can also be distributed to new areas on uncleaned farm implements and in manure from infected clover hay; steps should be taken to prevent this. Where clover rot is prevalent heavy dressings of nitrogenous fertilizer should be avoided. Grazing or cutting in autumn may be beneficial by reducing the amount of foliage available for ascospore infection and by reducing humidity in the crop which favours infection, but excessive defoliation then may cause more damage than the disease itself. There are no firm recommendations for chemical control, but quintozene or benomyl applied to the soil can suppress apothecial production from the sclerotia and hence reduce disease incidence. Some of the newer tetraploid cultivars of red clover show considerable resistance, as do some white clover cultivars. The NIAB lists give disease ratings to recommended cultivars of both, and these should be consulted.

The Fusarium root rot complex (*Fusarium* spp. with other fungi). Many *Fusarium* spp. are involved, including *F. oxysporum, F. roseum, F. solani, F. moniliforme* and *F. avenaceum*, along with weakly pathogenic fungi such as *Rhizoctonia* and *Pythium* spp. and even the normally saprophytic *Alternaria, Trichoderma* and *Rhizopus* spp. These fungi, with only quantitative variations, are found in the roots of all the common herbage legumes, including red and white clover and lucerne, from a few days after germination and thereafter throughout the life of the plant. The root rot symptoms appear as a progressive necrotic breakdown of the tap root and lateral roots, when the plant becomes subjected to some form of stress: too frequent cutting, extreme conditions of cold or water deficit, major nutrient deficiency (particularly potash), insect injury to the roots, or infection with leaf spotting fungi, nematodes or viruses; then these root-invading pathogens which may have lain dormant become active and increase in quantity.

No ready solutions to the problem have been found, other than management to reduce these stresses where possible. Experimentally, benomyl has been found to delay root rot development in red clover and hence to increase both yield and persistency.

Stem and leaf diseases

Clover scorch. Clover scorch (*Kabatiella caulivora*), commonly called 'northern anthracnose' in the USA, is one of the major diseases affecting red clover in temperate climates (Plate VIIb). It can also infect white clover but is unimportant there. Affected red clover fields show a blackening and breaking of the stems and withering of the leaves as if the whole crop were scorched. A wet season and high humidity within the crop favours infection, when temporary severe loss of foliage and, in a seed crop, a total failure of seed production, may occur. Lesions appear in late spring, up to 50 mm long, light to dark brown and sunken with dark margins on the stems and petioles; smaller and round on the leaves. Those on stems and petioles deepen until they penetrate to the pith and even girdle the whole stem, so that leaves and inflorescences, lacking water, wither and hang down. The fungus sporulates profusely on all infected parts; spores are spread by rain splash to other plants. The disease is seed borne (19) and the fungus can also over-winter on infected plants and decayed leaves, sporulating in the following spring to repeat the infection process.

Eradication from seed can be achieved with captan or organomercurial seed treatments, but there are no firm recommendations. An interval of several years is required between red clover crops to avoid recurrence of the trouble. Tetraploid are said to be more resistant than diploid cultivars and the US cultivar Lakeland has good resistance, but no active resistance breeding programmes are currently being pursued in Britain.

Powdery mildew. Powdery mildew (*Erysiphe trifolii*) can reach quite serious proportions on red clover from mid-summer onwards, particularly in warm, dry seasons. It can cause severe infection of sainfoin and also infects white clover, but not seriously.

The first sign of an attack is often a yellow mottling, then a superficial weft of whitish grey mycelium appears on the upper surfaces. From this prostrate mycelium arise erect conidiophores abstricting colourless conidia, which give the mildewed areas their powdery appearance, become detached and are carried by air currents to infect other plants.

Leaves which have been infected for some time tend to become chlorotic and finally dry and brown. By this stage, generally late in the season, small globose cleistothecia, initially light straw coloured but darkening rapidly as they mature, can be seen within the mycelial weft. These contain ascospores which are discharged forcibly and disseminated by the wind.

Tridemorph has proved effective in controlling the disase in small scale trials but has not been tested on a large scale.

Downy mildew.
This disease (*Peronospora trifoliorum*) is found occasionally on red clover but is commoner on white clover, especially following a wet spring (Plate VIIc). The first signs of infection are light, grey-green patches on the leaves. These increase in size and become puckered, while at the same time the undersides of the leaves become covered with a purplish grey weft of mycelium, from which sporangiophores arise in tufts. They are dichotomously branched and bear at their pointed tips solitary, oval to round, sporangiospores which are carried by the wind to other plants. Over-wintering is accomplished in systemically infected plants which have survived the winter and produce a fresh crop of conidia in spring. Downy mildew may also be seed borne.

Pseudopeziza leaf spot of clovers.
This leaf spot (*Pseudopeziza trifolii*) is a serious problem on red and white clover, where separate forms of *P. trifolii* will not infect the other's host. In addition to reducing quality, crop vigour and regrowth potential, the disease also increases oestrogen content in the legume, which may affect animal reproductive physiology (25).

Circular, brown to almost black spots up to 3 mm across occur on both leaf surfaces, with more elongate ones which seldom fruit on stems and petioles. A fruiting body (apothecium) is formed in the centre of each spot. This is light coloured and jelly like when expanded in wet weather to display the slightly convex upper surface, but in dry weather it appears darker when the margins are inrolled. Infection appears to be entirely by ascospores forcibly ejected from these apothecia and carried on the wind. A fresh crop of apothecia formed on overwintered leaves serves to start the infection going again in spring.

It is advisable to cut the crop early where a bad attack might otherwise cause premature leaf shed as well as leading to a build up of disease inoculum to infect neighbouring crops. In the USA, benomyl, zinc/maneb and chlorothalonil sprays have all proved effective in controlling the disease on lucerne but there are no specific UK recommendations.

Rusts.
Rusts (*Uromyces* spp.) are not such serious pathogens in herbage legumes as are those *Puccinia* spp. infecting grasses. Nonetheless they can cause damage on some occasions and there is evidence, as with some other herbage legume diseases, of increased oestrogen levels in infected plants which might cause reproductive disorders in farm animals (25).

U. fallens (included in older literature in *U. trifolii*) infects red and crimson clover and is often locally abundant. Small, scattered pale brown pustules (uredosori) cover the leaves in mid-season, followed by the less abundant, darker teleutosori. White and alsike clovers are similarly infected, though more rarely, with *U. trifolii* (referred to by some as *U. trifolii-repentis*). A commoner rust on white clover is *U. nerviphilus*, which forms reddish brown pustules mainly on veins and petioles where they cause swellings up to several mm long and marked distortion. This rust has been found causing serious damage and even death to breeders' material in the glasshouse and less commonly in field plots. No control measures are known, or deemed necessary, for any of these rust diseases.

Burn. Burn (*Leptosphaerulina trifolii*), or pepper spot, is one of the commonest diseases on white clover, reducing both yield and quality, especially crude protein content. It also raises the oestrogen content of the leaves of its legume host with possible effects on animal reproduction (17). The disease is prevalent in wetter areas from midsummer onwards, causing numerous, discrete small dark brown spots varying in size from well under 1 mm to 3 mm across, on both sides of the leaves but also on the petioles and even the pedicels. Badly attacked (more than 50% infected) leaves shrivel up, presenting a burnt appearance. The fruiting bodies (perithecia) are immersed in the larger lesions, which show light brown centres and dark brown boundaries. Ascospores are discharged forcibly from these perithecia and are disseminated by wind and rain, moisture being required for infection. Those from perithecia overwintering on dead leaves can start the spring infection going. The fungus is said to be seed borne. No control measures are known.

Diseases Caused by Viruses

Some 15 viruses infect herbage legumes in Britain (4). Many occur also in other legumes, peas, field and french beans for example. Natural transmission from the perennial herbage legumes to these annual field crops has been well documented in a number of cases. Some have hosts outside the legumes as well, nearly all dicotyledonous plants. Many of the same principles apply here as with the grass viruses (p. 68), especially the difficulty of diagnosis based exclusively on symptoms. Probably rather less is known than with grass viruses about their effects on the crop, as most surveys of herbage legume viruses have been qualitative rather than quantitative. What has become apparent, however, is that infection is commonly by more than one virus. For example, work at the Welsh Plant Breeding Station has shown that in a random sample of 68 white clover plants from upland sites 33 were infected with two, three with three and two with four viruses; only four plants were virus free. The frequency of viruses isolated was: white clover mosaic (53 plants), clover yellow vein (51), red clover vein mosaic (four) and bean yellow mosaic (four).

It could be that multiple infection is of greater importance to yield and persistence than that of any one virus in isolation. This is suggested by work in the USA where bean yellow mosaic and alfalfa mosaic viruses in combination reduced the herbage yield of Ladino white clover by 23–55% and flowering by 30%. Similar work in Canada showed 43% lower white clover dry-matter yields with a combined infection of white clover mosaic, clover yellow mosaic and alfalfa mosaic viruses, while in Scandinavia they were 21% lower with a double infection of white clover mosaic and clover yellow mosaic viruses.

It is difficult to estimate what effect infection of the legume companion might have on balance in grass/legume swards, but clearly the reduction it represents in the amount of fixed nitrogen available to the grass must reduce overall productivity. There is indeed some experimental evidence to this effect. In the absence of compensation by the legume constituent any thinning of the grass must lead to weed encroachment.

Only the six most commonly identified and important viruses of herbage legumes are considered here.

Red clover necrotic mosaic virus (RCNMV). (Plate VIId.) RCNMV was discovered in Europe in 1967 and appeared in Britain in 1973. Soon after, it was found causing massive damage to the red clover cultivar, Hungaropoly, at various trial sites and has since appeared in commercial crops. In red clover it causes severe leaf mottle and distortion, vein and more general necrosis and moderate to severe stunting during winter and early spring, but if the plants survive, the symptoms tend to fade over the summer months. The virus has been isolated from leaves, stems and roots; it is readily sap transmissible and transmission through the soil has been demonstrated. Seed transmission has been suspected, but tests have failed to confirm this. Expert diagnosis is essential, particularly as some isolates of white clover mosaic virus induce somewhat similar symptoms.

Prospects of amelioration through the use of resistant cultivars appear promising.

Alfalfa (lucerne) mosaic virus (AMV). This virus is one of several causing mosaics in red and white clover. French beans and peas are infected naturally, and in all AMV has some 150 hosts in 22 families of dicotyledons. The virus can reduce the yield of lucerne by 20% and increase predisposition to winter injury and drought. Its importance for clovers in Britain is less clear. AMV was not isolated from clover pastures in a survey carried out some 15 years ago but it is widespread elsewhere and in the USA infection of Ladino white clover with AMV and bean yellow mosaic virus reduced yields by 25–55%.

The virus is transmitted readily by infected sap and also by a large number of aphids in the non-persistent manner, among them the pea aphid (*Acyrthosiphon pisum*), the peach-potato aphid (*Myzus persicae*) and the black bean aphid (*Aphis fabae*). It is also seed borne; levels of up to 10% seed transmission have been reported for commercial stocks and up to 50% in seeds from individual infected plants.

White clover mosaic virus (WCMV). WCMV is common in many parts of the temperate world. In Britain it has been found occurring naturally in white, red, alsike and crimson clovers and to a lesser extent in lucerne and peas. The commonest symptoms in clovers are a light green striping or flecking of the leaves between the veins. White clover can be symptomless, but transfer of virus from such symptomless plants to red clover often results in a mild, necrotic mottle. Sap from infected plants is highly infectious. The pea aphids (*Acyrthosiphon pisum* and *Macrosiphum pisi*) may be vectors, but this is not certain.

Work at Aberystwyth has shown that WCMV reduced the dry-matter yield of several white clover cultivars on the NIAB recommended list by an average of 10%, losses for individual cultivars ranging from 3–18%. The greatest losses were shown by the cultivars Milkanova and Menna, and the least by Olwen, S.100 and S.184. Seed yields and the numbers of inflorescences, florets per head and seed size were also reduced and cultivar responses followed the same pattern as for reduction in herbage yield. Similar effects have been noted in red clover infected with this virus, which also adversely influences persistence and increases winter kill.

Bean yellow mosaic virus (BYMV). BYMV is confined mainly to legumes, in many of which it causes mosaic symptoms. The well known pea mosaic is a variant of this virus; another causes necrosis in red clover, from which the virus has been isolated quite frequently in Britain. Infected white clover is generally symptomless and BYMV has been isolated only occasionally from white clover pastures in Britain though in the USA BYMV in combination with lucerne mosaic virus (q.v.) caused losses of 25–55% in Ladino white clover. It is transmissible by infected sap and also by some 20 aphid species, the commonest being the pea aphid (*Acyrthosiphon pisum*), the potato aphid (*Macrosiphum euphorbiae*), the peach-potato aphid (*Myzus persicae*) and the black bean aphid (*Aphis fabae*). The chief importance of this virus probably lies in the perennial reservoir provided by red clover in particular for infection of the pea crop, where it is particularly serious.

Clover yellow vein virus (CYVV). This virus, related to bean yellow mosaic (q.v.) causes a mild, veinal yellowing and mottle in white clover, in which it has been found quite commonly in Britain, often in combination with white clover mosaic virus (q.v.). Red clover is less frequently affected and symptoms are rare in this host. Peas and beans are not generally infected. The virus is sap and aphid transmissible in the non-persistent manner by the peach-potato aphid (*Myzus persicae*), the pea aphid (*Acyrthosiphon pisum*) and the potato aphid (*Macrosiphum euphorbiae*) but not by the black bean aphid (*Aphis fabae*). Nothing of substance is known concerning its effect on the clover crop.

Red clover vein mosaic virus (RCVMV). Most hosts of RCVMV are legumes, and this is one of the few viruses in which symptoms in clovers are fairly characteristic: marked vein yellowing with only mild, inconspicuous intervenal mottling. These are most pronounced in red clover, particularly in late spring; infected white clover has similar though less conspicuous symptoms which may be absent altogether. Earlier surveys found this to be one of the commonest viruses in white and red clover pastures, although it was less common in white clover from upland pastures. The virus is transmitted through the seed in red clover. It is also readily sap transmissible and can be transmitted by aphids in the non-persistent manner, the commonest vector being the pea aphid (*Acyrthosiphon pisum*). Nothing is known to date of its effect on herbage yields, but the common presence of RCVMV in clovers must represent a threat to the pea crop, where its effect is severe.

Control of herbage legume viruses

There seems little prospect for control of these viruses directly other than by resistance breeding where merited, of which there appears to be some prospect. Priorities would need to be determined in the light of information yet to be gained on the precise effects of these viruses on herbage yield. Indirect control might be feasible through destruction of the aphid vectors by appropriate insecticides, though as explained earlier because aphids must feed on sprayed plants in order to be killed they also put virus in; control would thus not be complete though virus levels would be reduced. Care would need to be taken to avoid harmful effects to stock, and to avoid killing bees. Granule treatments or pirimicarb sprays are less toxic to bees, and during flowering sprays should be applied in the evening after honey bees have ceased flying. It is doubtful whether sprays on herbage legumes as constituents of pastures for grazing or conservation would be economically justifiable, and no firm recommendations are given. However, they could be of value in seed crops, especially where the aphids listed along with the vetch aphid (*Megoura viciae*) and the black legume aphid (*Aphis craccivora*) are sufficiently abundant to become pests in their own right. Effective aphicides include: *granules:* disulfoton (850 g ha^{-1}), phorate (1.1 kg ha^{-1}); *sprays* (used at least at 225 l ha^{-1}): demephion (250 g ha^{-1}), demeton-S-methyl (245 g ha^{-1}), oxydemeton-methyl (240 g ha^{-1}), pirimicarb (140 g ha^{-1}) and thiometon (275 g ha^{-1}). All are eradicant treatments; all except pirimicarb and thiometon also have a preventive action.

Diseases Caused by Mycoplasma

Three diseases, thought to be distinct, are known in clovers: phyllody, witches' broom and clover red leaf (English stolbur), which were at one time thought to be caused by viruses. Symptoms of these 'yellows type' diseases are also somewhat virus-like: chlorosis, morphological abnormalities such as excess axillary proliferation (witches' brooms) or floral phyllody, which are probably the result of induced hormonal imbalance in the hosts. However, no virus particles were observed either in intact plants or in expressed sap.

In 1967 it became clear that the causal organisms are mycoplasma-like organisms (MLOs) present in the phloem elements of infected plants. Deep feeding leafhoppers acquire these MLOs from the phloem of infected plants, there is a long latent period of several weeks while the organism multiplies in the body of the vector before it becomes infective, but thereafter the vector is able to transmit during the whole of its natural life and may even pass the infective principle to its offspring.

Clover phyllody. Phyllody affects most clovers, particularly white clover (Plate VIIe). It is transmitted by various leafhoppers, principally *Euscelis plebeja, Macrosteles* spp. and *Aphrodes bicinctus*. The same MLO also causes green petal disease in strawberries and infects many annual weeds. Affected clovers become rather stunted, with vein clearing, general chlorosis and bronzing of the older leaves. There is some axillary proliferation leading to the production of a number of weak looking shoots. The most characteristic symptom is the transformation of the inflorescences into leafy structures which set no seed. Surveys in Britain established that between 1959 and 1962 40% of white clover seed crops were infected, levels within crops varying from a trace to 35%. Prevalence was greatest in older crops and greater in large leaved than in wild white cultivars. The disease was rarer in red clover stands. Since that time incidence has fluctuated considerably. A series of cold winters led to a declining leafhopper population but there is evidence of a recent increase. In plot experiments white clover vigour was markedly affected, this effect not being compensated for entirely by increased grass growth, probably because of the known depressing effect on nodulation and hence nitrogen fixation, with the result that there was considerable encroachment by broad-leaved weeds.

Control of phyllody on a field scale is impracticable. Although insecticides will kill the leafhoppers, most spread, where there is a long latent period in the vector, is from outside the crop, and insecticides are more effective in preventing spread within rather than reducing external sources. No resistant white clovers have been discovered on which to base a breeding programme. Valuable breeding material which has become infected can, however, be saved sufficiently to enable seed to be taken from it by dry heat treatment (7–10 days at 40°), ultraviolet irradiation, culture in nutrient solutions containing high levels of zinc salts (zinc sulphate at 10–20 ppm), and by treatment with tetracycline antibiotics. Rarely do these effect a permanent 'cure'; generally the plants slowly revert to the phyllody condition.

Clover witches' broom. This disease, caused by the same MLO which induces bronze leaf wilt in strawberries, is transmitted by several leaf hoppers which also transmit clover phyllody, but not by *Aphrodes bicinctus*. Clover plants become markedly dwarfed, show a high degree of crown and axillary proliferation leading to pronounced tufts of weak shoots (brooms) and the leaves are small with chlorotic and eventually bronzed leaf margins. Delay in unfolding of the leaflets gives rise to a 'club leaf' condition of the younger leaves. Flowering is partially suppressed but those inflorescences that are produced do not display floral phyllody. White clover yield is reduced more by witches' broom than by phyllody but the former appears less common in swards. There is no method of control.

Clover red leaf (English stolbur). (Plate VIIf.) Stolbur type infections are well known, causing serious damage to crops such as potatoes, tomatoes and tobacco. The stolbur type disease of clovers commonly referred to as 'red leaf' is widely distributed in white and not unknown in red clover. In white clover it causes an initial marginal leaf chlorosis followed by extreme bronzing of the foliage. The plants become increasingly stunted and eventually die. Some axillary proliferation occurs but this is not as marked as with witches' broom, while flowering is often reduced but the inflorescences, although small, are normal in appearance and seed set. Clover red leaf is transmitted by *Euscelis plebeja* and *Aphrodes bicinctus* but not by *Macrosteles sexnotatus* which transmits the other two mycoplasmal diseases. No methods of control are known.

References

1. A'Brook, J.; Heard, A. J. (1975). The effect of ryegrass mosaic virus on the yield of perennial ryegrass swards. *Annals of Applied Biology,* **80,** 163–168.
2. Aldrich, D. T. A.; Doling, D. A. (1967). Varietal resistance to clover rot in white clover. *Nature, London,* **214,** 946–947.
3. Anon. (1971). *Clover Rot.* Ministry of Agriculture, Fisheries and Food Advisory Leaflet 266, H.M.S.O., 4 pp.
4. Carr, A. J. H. (1971). Herbage legume diseases; Virus diseases of forage legumes; grass diseases. In *Diseases of crop plants,* pp. 254–307, Western, J. H. (Ed.), London: MacMillan, 404 pp.
5. Catherall, P. L. (1966). The significance of virus diseases for the productivity of grassland. *Journal of the British Grassland Society,* **21,** 116–122.
6. Catherall, P. L. (1979). Virus diseases of cereals and grasses and their control through plant breeding. *Welsh Plant Breeding Station Annual Report for 1978* 205–226.
7. Catherall, P. L. (1981). *Virus diseases of grasses.* Ministry of Agriculture, Fisheries and Food Leaflet 595, H.M.S.O., 8 pp.
8. Catherall, P. L.; Wilkins, P. W. (1977a). Some problems and progress in breeding herbage grasses for virus resistance. *Annales de Phytopathologie,* **9,** 245–248.
9. Catherall, P. L.; Wilkins, P. W. (1977b). Barley yellow dwarf virus in relation to the breeding and assessment of herbage grasses for yield and uniformity. *Euphytica,* **26,** 385–391.
10. Davies, H.; Williams, A. E.; Morgan, W. A. (1970). The effect of mildew and leaf blotch on both yield and quality of *Lolium multiflorum* (cv. Lior). *Plant Pathology,* **19,** 135–137.
11. Dickens, J. S. W.; Mantle, P. G. (1974). *Ergot of cereals and grasses.* Ministry of Agriculture, Fisheries and Food Advisory Leaflet 548. H.M.S.O., 3 pp.
12. Egli, Th.; Goto, M.; Schmidt, D. (1975). Bacterial wilt, a new forage grass disease. *Phytopathologische Zeitschrift,* **82,** 111–121.
13. Gray, E. G.; Copeman, G. J. F. (1975). The role of snow moulds in winter damage to grassland in northern Scotland. *Annals of Applied Biology,* **81,** 235–279.
14. Lancashire, J. A.; Latch, G. C. M. (1970). The effect of crown rust (*Puccinia coronata* Corda) on the yield and botanical composition of two ryegrass/white clover pastures. *New Zealand Journal of Agricultural Research,* **13,** 279–286.
15. McGee, D. C. (1971). The effect of benomyl on *Gloeotinia temulenta* under laboratory and field conditions. *Australian Journal of Experimental Agriculture and Animal Husbandry,* **11,** 693–695.
16. Michail, S. H.; Carr, A. J. H. (1966). Italian ryegrass, a new host for *Ligniera junci. Transactions of the British Mycological Society,* **49,** 411–418.
17. Newton, J. E.; Betts, J. E.; Drane, H. M.; Saba, N. (1970). The oestrogenic activity of white clover. In *White clover research.* British Grassland Society Occasional Symposium, **6,** 309–314.

18. Noble, M.; Richardson, M. J. (1968). *An annotated list of seed-borne diseases.* Commonwealth Mycological Institute, Kew, Surrey, Phytopathological Papers No. 8, 191 pp.
19. Nüesch, B. (1964). "Renova" und "Lior", zwei neue Futter Pflanzensorten. *Mitteilungen für die Schweizerische Landwirtschaft, Zurich,* **12,** 1–12.
20. O'Rourke, C. J. (1976). *Diseases of grasses and forage legumes in Ireland.* Carlow, Eire: The Agricultural Institute, Oak Park Research Centre, 115 pp.
21. Wilkins, P. W. (1973a). Infection of *Lolium multiflorum* with *Rhynchosporium* species. *Plant Pathology,* **22,** 107-111.
22. Wilkins, P. W. (1973b). *Puccinia recondita* on ryegrass. *Plant Pathology,* **22,** 198.
23. Wilkins, P. W. (1973c). Infection of *Lolium* and *Festuca* spp. by *Drechslera siccans* and *D. catenaria. Euphytica,* **22,** 106–113.
24. Wilkins, P. W.; Exley, J. K. (1977). Bacterial wilt of ryegrass in Britain. *Plant Pathology,* **26,** 99.
25. Wong, E., Flux, D. S.; Latch, G. C. M. (1971). The oestrogenic activity of white clover (*Trifolium repens* L.). *New Zealand Journal of Agricultural Research,* **14,** 639–645.
26. Wright, C. E. (1967). Blind seed disease of ryegrass. *Euphytica,* **16,** 122–130.

A further useful publication, containing excellent coloured illustrations, is:

Priestley, R. H.; Bayles, Rosemary A. (1982). *Identification and control of cereal diseases.* Cambridge: National Institute of Agricultural Botany, 41 pp.

Appendix I   IDENTIFICATION OF BROAD-LEAVED WEEDS AND CROP AND WEED GRASSES

Because of the importance of early weed control, it is necessary to recognize which weeds are present as soon as possible. Broad-leaved weeds have specific diagnostic features, as indicated in Table A I.1. They are usually easy to see and identify.

Grasses, on the other hand, are much more difficult to recognize, especially at the seedling stage when only one or two leaves are present; often it is necessary to wait until several leaves have expanded fully.

TABLE A I.1. Diagnostic features of some common broad-leaved weeds found in newly sown leys

Species	Germination	Cotyledons	Leaves	Young plant habit	Stems	Flowers
Charlock (*Sinapis arvensis*)	Spring, autumn	Kidney-shaped	Broad, rough-hairy	Tall	Thick	Bright yellow
Chickweed, Common (*Stellaria media*)	All months	Small	Small, oval, paired	Light green clumps	Trailing	Small, white
Corn spurrey (*Spergula arvensis*)	Spring, autumn	Small narrow	Long, narrow, curved	Spider-like	Sticky	White, star-like
Fat-hen (*Chenopodium album*)	Spring, summer	Narrow, purple underneath	Silvery, coarsely tufted	Tall	Reddish	Light green inconspicuous
Fumitory (*Fumaria officinalis*)	Spring, early summer	Long, narrow	Deeply divided	Pale/green branched	Pink/green	Pink
Groundsel (*Senecio vulgaris*)	All months	Purple underneath	Step-like teeth	Erect, variable	Weak, fleshy	Yellow
Knot-grass (*Polygonum aviculare*)	Spring only	Narrow, point upwards	Dark green hairless	Vertical (intense crops)	Tough	Pink/White
Mayweed, Scentless (*Matricaria perforata*)	All months	Small	Narrow, divided, dark green, aromatic	Wiry tufts	Branched	White/yellow
Shepherd's Purse (*Capsella bursa-pastoris*)	All months	Small	Deeply lobed, grey/green	Small rosette	Tough	White
Speedwell, Common Field (*Veronica persica*)	All months	Rounded triangular	Hairy, notched	Short, sprawling	Weak	Large, blue/white

One of the key characters for identification is the ligule, a thin membranous outgrowth between the leaf blade and the leaf sheath (Fig. A I.1). It varies greatly in shape and length between species, although confusing plant-to-plant variations within the same species are often found. Another important diagnostic feature is the auricle, or enlarged leaf base. However, these two characters are seldom enough for sure identification; other associated characteristics can include the way the youngest leaf is enclosed in the older leaf sheath (i.e. whether folded or rolled), morphology and shape of leaves, and the extent of hairiness (Fig. A I.2).

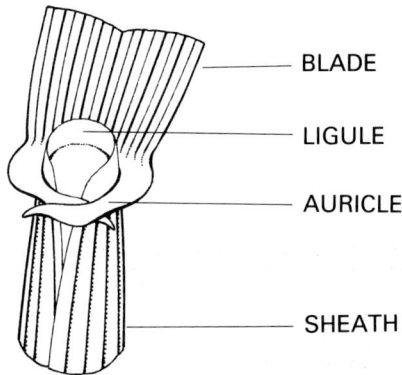

BLADE

LIGULE

AURICLE

SHEATH

Fig. A I.1   Close-up of leaf base of *Elymus (Agropyron) repens*.

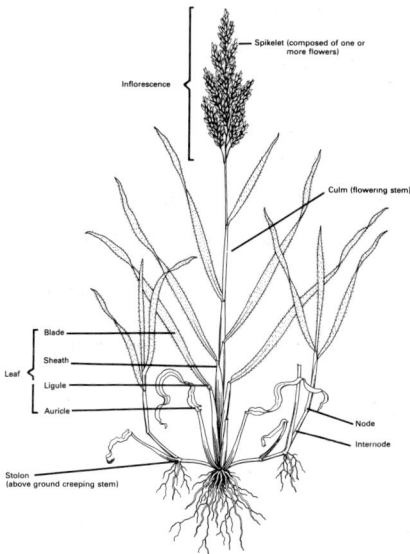

Fig. A I.2   Diagram of a grass plant showing diagnostic characters.

A suggested key for identifying young grasses, before they flower, is given in Table A I.2. This key includes a short descriptive feature for each species. Note that, under practical conditions, the plants are likely to be partially defoliated, or their growth habit modified (e.g. etiolated) by the companion species and weeds. Consequently, not all the key features may be noticeable – in which case check that *most* features are present.

TABLE A I.2.   Key for Identification of Vegetative Grasses

GROUP A			
Leaves Bristle-like			
Species	Plant Characteristics	Leaves	Ligules
Wavy hair-grass (*Deschampsia flexuosa*)	Plant rather stiff	Long, dark-green	Short, broad
Mat-grass (*Nardus stricta*)	Shoots bunched; tough, shiny sheaths	Outer horizontal	Short, blunt
Sheep's fescue (*Festuca ovina*)	Leaf sheath open/split; occasionally pink based	Fine, thin	Minute
Red fescue (*Festuca rubra*)	Leaf sheath not split initially; red sheaths; rhizomes	Infolded	Minute
GROUP B			
Leaves Folded			
Species	Plant Characteristics	Leaves	Ligules
Perennial ryegrass (*Lolium perenne*)	Shiny, dark green; red basal sheath	Shiny below	Small, blunt
Cocksfoot (*Dactylis glomerata*)	Large succulent base	Broad, puckered	Pronounced
Annual meadow-grass (*Poa annua*) (Plate Ia,b)	Tufted; sheaths green-white	Short, crinkled	Large, white
Rough meadow-grass (*Poa trivialis*) (Plate Ic)	Curved tillers; sheaths brownish; stolons; often rough stemmed	Shiny below	Uppers long, pointed
Smooth meadow-grass (*Poa pratensis*)	Dark green; smooth stemmed; rhizomes	Parallel-sided	Short
Heath grass (*Danthonia decumbens*)	Hairs at leaf base	Tramlined	Short hairs
Crested dogstail (*Cynosurus cristatus*)	Leaf sheath sometimes yellowish	Trough-like, finely pointed	Short wrap-around

TABLE A I.2 *Continued*

GROUP C
Leaves Rolled; Plant Hairy; Auricles Present

Species	Plant Characteristics	Leaves	Ligules
Common Couch (*Elymus* (*Agropyron*) *repens*) (Plate Ie,f)	Strong rhizomes; prominent auricles	Long, dull green, downy above	Very short
Wall Barley (*Hordeum murinum*) (Plate Ig,h)	Downy annual	Thin, hairy both sides	Short
Meadow barley (*Hordeum secalinum*)	Downy perennial	Hairy but not obvious below	Very short
Wheat (*Triticum* spp.)	Auricles fringed	Large, tapering	Large

GROUP D
Leaves Rolled; Plant Hairy; Auricles Absent

Species	Plant Characteristics	Leaves	Ligules
Barren brome (*Bromus sterilis*) (Plate IIa,b)	Annual; weak stems	Soft blades, twisted twice	Long, ragged
Soft brome (*Bromus hordeaceus*) (*mollis*) (Plate IIc,d)	Tufted, hairy	Thin, flaccid	Short, serrated
Yorkshire fog (*Holcus lanatus*) (Plate IIe,f)	Tufted, velvety; pink veins on sheaths	Hairy, grey-green	Large, toothed
Creeping soft-grass (*Holcus mollis*) (Plate IIg,h)	Large rhizomes; long hairs on nodes	Softly hairy	Serrated, blunt
False Oat-grass (*Arrhenatherum elatius*) (Plate IIIa,b,c)	Tall; often onion-based	Long, pointed	Short, round
Sweet Vernal grass (*Anthoxanthum odoratum*) (Plate IIId: *A. puelii*)	Spreading hairs at sheath blade junction	Aromatic	Long; hairs

GROUP E
Leaves Rolled; Plant Glabrous (hairless); Auricles Present

Species	Plant Characteristics	Leaves	Ligules
Italian ryegrass (*Lolium multiflorum*)	Bright red base	Dark green; shiny beneath	Short, blunt
Meadow fescue (*Festuca pratensis*)	Dark brown basal sheaths	Stiffly erect; spiky	Small, blunt
Tall fescue (*Festuca arundinacea*)	Coarse habit; auricles fringed	Broad, dark-green	Short
Barley (*Hordeum* spp.)	Big clasping auricles	Large, pointed	Small

TABLE A I.2. *Continued*

GROUP F
Leaves Rolled; Plant Glabrous (hairless); Auricles Absent

Species	Plant Characteristics	Leaves	Ligules
Timothy (*Phleum pratense*) (Plate IIIe,f)	Stem base sometimes swollen	Large, pale green	Long, white
Common Bent (*Agrostis capillaris*) (*tenuis*) (Plate IIIg,h)	Dense; erect; rhizomes cordlike	Triangular, short	Short, blunt
Creeping Bent (*Agrostis stolonifera*) (Plate IVb,c)	Small plant; stolons	Triangular, small	Long, rounded
Black Bent (*Agrostis gigantea*) (Plate IVd,f)	Large plant, rhizomes	Triangular, long	Long, ragged
Meadow foxtail (*Alopecurus pratensis*)	Tufted perennial	Smooth, pale green	Small, blunt
Black-grass (*Alopecurus myosuroides*) (Plate IVa)	Tufted annual	Veins large	Long, white
Tufted hair-grass (*Deschampsia caespitosa*)	Tall; tussock; harsh to touch	Dark green; rough (ribbed)	Long, pointed
Crested Dogstail (*Cynosurus cristatus*)	Leaf sheaths sometimes yellowish	Trough-like, finely pointed	Short, wrap-around
Oats (*Avena* spp.) (Plate Id, IVe)	Annual; basal leaves loosely hairy	Broad, light, blue-green	Long, blunt

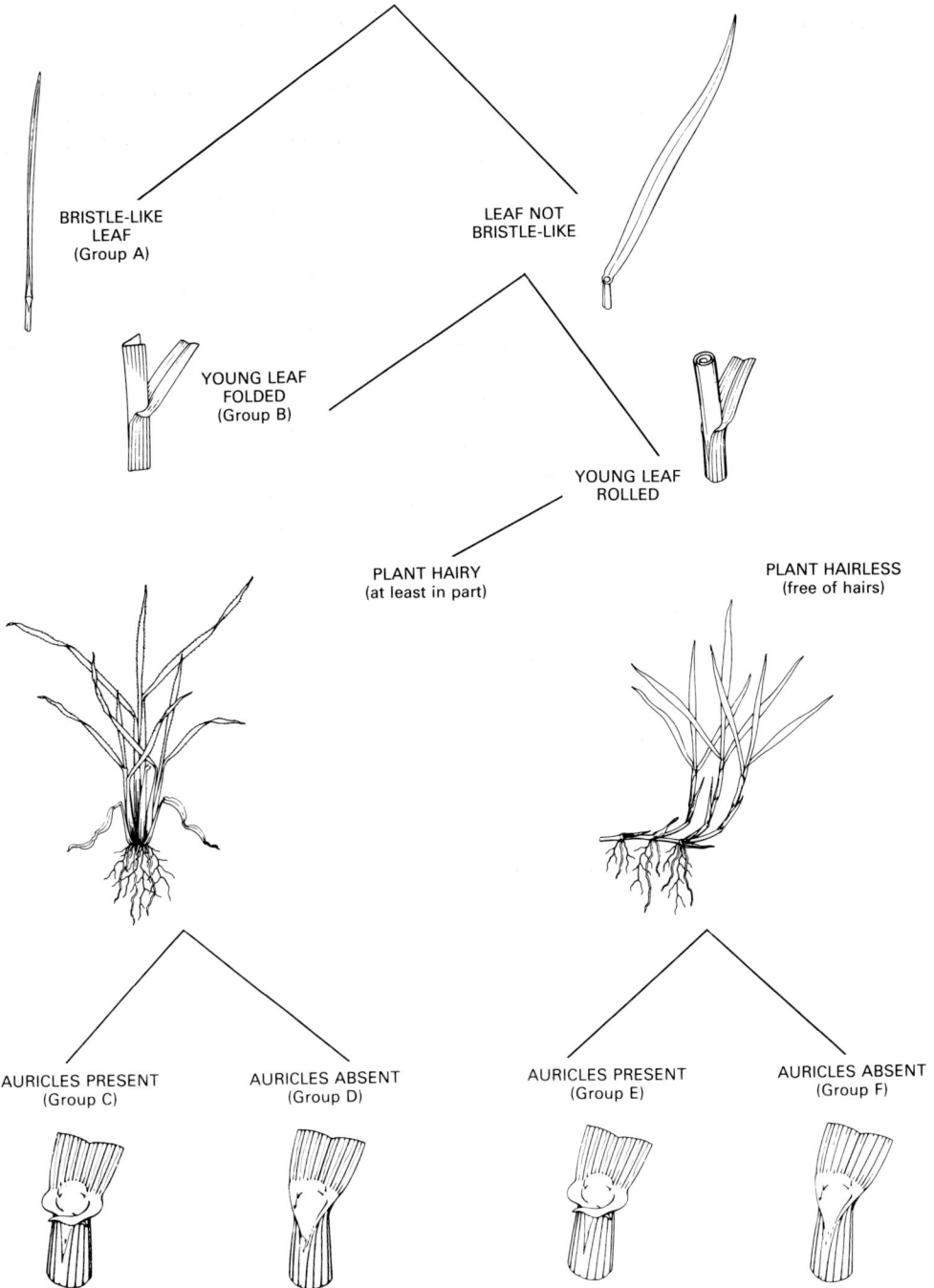

Fig. A I.3 Identification of grasses in the vegetative state.

Appendix II   SAFETY PRECAUTIONS

Most, if not all, of the preparations used as herbicides or pesticides are toxic to other forms of life, including people, even if they are described as selective or specific. It is therefore essential that proper precautions be taken in their use, in order to avoid injury to operators and others and damage to the environment by killing wildlife or upsetting its balance.

There is now a large volume of legislation, as can be seen from the list given below of relevant Acts; the simple rule to follow when using any herbicide, pesticide or fungicide is to use only an approved product and to follow the instructions on the label implicitly. An 'approved product' means one approved under the Agricultural Chemicals Approval Scheme: although participation in this scheme is voluntary, it is supported by the various associations of manufacturers, suppliers and users, and preparations which are not approved are unlikely to be available on the market. An approved product is granted a certificate of approval which is renewable annually, and its label carries an easily recognizable identification of this (a large 'A' surmounted by a crown, over the words Agricultural Chemicals Approval Scheme). An annual list of approved products *Approved Products for Farmers & Growers* is published in February and is available from bookshops or from the Ministry of Agriculture, Fisheries and Food (Publications), Lion House, Willowburn Trading Estate, Alnwick, Northumberland NE66 2PF.

More detailed consideration of the scheme, and of the Pesticides Safety Precautions Scheme and the various Acts and Regulations may be found in the *Pest and Disease Control Handbook* (2nd edition, 1983, Nigel Scopes and Michael Ledieu, Eds.) and the *Weed Control Handbook: Principles* (7th edition, 1982, H. A. Roberts, Ed.) both published by the British Crop Protection Council.

Relevant Acts and Regulations
Health and Safety at Work etc. Act 1974
Health and Safety (Agriculture) (Poisonous Substances) Regulations 1975
Poisons Act 1972
Rivers (Prevention of Pollution) Acts 1951 and 1961
Deposit of Poisonous Wastes Act 1972
Control of Pollution Act 1974
Wildlife & Countryside Act 1981

INDEX

Abacarus hystrix, RMV transmission, 69
Acyrthosiphon pisum — *see* Aphids, Pea aphid
Agriolimax reticulatum — *see Deroceras reticulatum*
Agriotes spp. — *see* Wireworms
Agropyron repens — *see Elymus repens*
Agrostis capillaris — *see* Bent, common
Agrostis capillaris tenuis — *see* Bent, common
Agrostis gigantea — *see* Bent, black
Agrostis stolonifera — *see* Bent, creeping
Agrostis spp. — *see* Bent grasses, Bents
Aldicarb, mites, 69
Aldrin, not permitted, 28
Alfalfa mosaic virus
　losses, 76
　transmission, 76
All-grass leys, broad-leaved weed control, 22 (table)
All-grass mixtures, seedling broad-leaved weed control, 19
Allium vineale — *see* Wild onion
Alopecurus myosuroides — *see* Black grass
Alopecurus pratensis — *see* Meadow foxtail
Alsike clover
　rust, 74
　WCMV, 76
Alternaria spp. — *see* Fusarium root rot complex
Altica carduorum, 9
AMV — *see* Alfalfa mosaic virus
Annual meadow grass, pl. I, 1, 9, 18
　competition from chickweed, 14
　control in newly-sown leys, 25
　density, 18
　dry conditions, 33
　effect of nitrogen, 8
　following sward damage, 33
　identification (key), 84
　in grazing, 36
　snow mould, 53
　urine scorch, 34
　water-logged soils, 33
　see also Meadow grasses
Annual weeds
　grazing, 18
　mowing, 18
Anthoxanthum odoratum — *see* Sweet vernal grass
Anthoxanthum puelii, pl. III
Anthracnose — *see* Northern anthracnose
Anthriscus sylvestris — *see* Cow parsley
Antler moth, control of mass migrations, 50
　description, larva and adult, 50
　in established grassland, 50
　life cycle, 50

Antonina graminis, see Rhodes grass scale
Aphids, pl. V, 2
　BYMV transmission, 69-70, 77
　CSV transmission, 71
　cutting, 51
　description, 50
　grazing, 51
　in established grassland, 50-51
　life cycle, 50
　RCVMV transmission, 77
　virus transmission, 51
Aphis craccivora — *see* Black legume aphid
Aphis fabae — *see* Black bean aphid
Aphrodes bicinctus
　clover phyllody transmission, 78
　clover red leaf transmission, 79
Army worm — *see* Antler moth
Arrhenatherum elatius — *see* False oat grass
Asulam, 9
　bracken, 40
　broad-leaved weeds, susceptibility 38 (table)
　docks, 43
　marsh horsetail, 44
　time to subsequent crops, 44
　ragwort, 46
　restricted use on hay fields, 43
　swards containing clovers, 37
Asulam + mecoprop + MCPA
　broad-leaved weeds, susceptibility, 38 (table)
　damage to clover, 44
　docks, 44
　grass-only swards, 37
　ragwort, 46
Atropa belladonna — *see* Deadly nightshade
Auricle, 83 (fig.)
Avena spp. — *see* Oats

Bacillus popilliae — *see* Milky disease
Bacterial wilt of ryegrass
　cutting, 68
　description, 67
　nitrogen, 68
　resistance, 68
　spread, 68
　water stress, 68
Baits
　leatherjackets, 27
Barley
　halo spot, 65
　identification (key), 85
Barley grass
　in all-grass swards, 39
　in grass/clover swards, 40
　management, 39

Barley yellow dwarf virus, pl. VI, 51, 69-70
  breeding for resistance, 70
  description, 70
  effects, 70
  losses, 70
  slot seeding, 70
  transmission by aphids, 69-70
  with crown rust, 4
Barren brome, pl. II
  identification (key), 85
Beans, CYVV, 77
Bean yellow mosaic virus, 76-77
  transmission, 77
  with AMV, losses, 76
Beetle for biological control, 9
Beetles, 2
Benazolin, 9
Benazolin + 2,4-DB + MCPA
  all-grass leys, 22
  broad-leaved weeds, susceptibility, 38
    (table)
  chickweed, 24
  creeping buttercups, 41
  grass/clover leys, 22
  docks, 43
  nettles, 45
  swards containing clover, 37
  thistles, 47
Benazolin mixtures
  black bindweed, 23
  chickweed, 23
  cleavers, 23
  knotgrass, 23
  redshank, 23
Benazolin salt, perennial ryegrass, amount
  tolerated, 21
Benodanil
  crown rust, 54
  stripe rust, 63
Benomyl
  blind seed disease of ryegrass, 66
  clover rot, 72
  ergots, 67
  flag smut, 66
  fusarium root rot, 73
  powdery mildew, 63
  pseudopeziza leaf spot, 74
  rhynchosporium leaf blotch, 64
  snow mould, 54
Benomyl + ethofumesate + phorate, 14
Bent, blind seed disease of ryegrass, 66
Bent, black, pl. IV
  identification (key), 86
Bent, common, pl. III
  identification (key), 86
  low N, 32
  P and K deficiency, 32

Bent, creeping, pl. IV
  identification (key), 86
Bent grasses, pH, 31
Bent wire grass, antler moth, 50
Bent — see also Bents
Bentazone + dichlorprop, all-grass leys, 22
Bentazone + MCPB + MCPA
  all-grass leys, 22
  chickweed, 24
  grass/clover leys, 22
Bentazone mixtures
  chickweed, 23
  cleavers, 23
  corn marigold, 23
  mayweed, 23
Bents, 1
  choke, 67
  sensitivity to asulam, 43
  snow mould, 53
  see also Bent
Benzoylprop-ethyl, wild oats, 25
Bibio marci — see St. Mark's fly
Bibionids
  description, 51
  life cycle, 51
Biological control
  agents, 8 (table)
  bracken, 9
  Cirsium arvense, 9
  definition, 7
  insect pests, 10
  pathogens for, 11
  weeds, 9
Bird-cherry aphid, BYDV transmission, 69
Black bean aphid
  AMV transmission, 76
  CYVV, 77
Black bindweed
  herbicides for control, 23
  susceptibility to herbicides, 22 (table)
Black grass, pl. IV, 19
  crown rust, 54
  identification (key), 86
Black legume aphid, 77
Blackberry aphid, BYDV transmission, 69
Blade, 83 (fig.)
Blind seed disease of ryegrass
  life cycle, 66
  losses, 66
  management, 66
Bracken, 1, 11, 39
  biological control, 9
  cutting, 35
  in grass/clover swards, 40
  in grazing, 36
  management, 40
  poisoning, 40

susceptibility to herbicides, 38 (table)
when to spray, 38
Broad-leaved dock, occurrence, 42
Bromoxynil + ioxynil
all-grass leys, 22
pre-emergence spraying, 19
seedling broad-leaved weeds, 19
undersown crops, 19
Bromoxynil + mecoprop
all-grass leys, 22
Bromoxynil mixtures
black bindweed, 23
corn marigold, 23
knot grass, 23
mayweed, 23
redshank, 23
Bromus hordaceus — see Soft brome
Bromus mollis — see Bromus hordaceus
Bromus sterilis — see Barren brome
Bronze leaf wilt, 78
Brown rust
description, 63
occurrence, 54
Buckthorn, crown rust, 54
Bulbous buttercup — *see* Buttercup, bulbous
Burn, description, 75
Burning/grazing pressure, and chemical control, 14
Buttercup, bulbous, 41
susceptibility to herbicides, 38 (table)
when to spray, 38
Buttercup, creeping, 41
in grass/clover swards, 41
susceptibility to herbicides, 38 (table)
Buttercup meadow, 41
susceptibility to herbicides, 38 (table)
Buttercups, 1, 18, 39
identification, 41
in all-grass swards, 41
in grass/clover swards , 41
in grazing, 36
management, 41
poisonousness, 41
susceptibility to herbicides, 22 (table)
sward defoliation, 34
BYDV — *see* Barley yellow dwarf virus
BYMV — *see* Bean yellow mosaic virus

Capsella bursa-pastoris — see Shepherd's
purse
Captan, clover scorch, 73
Carduus acanthoides — see Welted thistle
Carduus nutans, 9 — *see also* Musk thistle
Carduus spp., 9
Cat's ear, susceptibility to herbicides, 38 (table)
Cattle

bracken poisoning, 40
ergot disorders, 66
ragwort poisoning, 45
selectivity, 34
Cerapteryx graminis — see Antler moth,
Army worm
Cereal leaf beetle, CfMV, 71
Cereal stubbles, paraquat, 19
CfMV — *see* Cocksfoot mottle virus
Chafer grubs, pl. V, 49-50
Chafers
distribution, 50
in established grassland, 49-50
management, 50
— *see also* Cockchafer, Garden chafer
Charlock
diagnostic features, 82 (table)
susceptibility to herbicides, 22 (table)
Chemical control, 7
grassland diseases, 4
pathogens, 10
Chenopodium album — see Fat hen
Chickweed, 1, 9, 18, 19
choice of herbicide, 23
conditions favouring, 23
control in newly sown leys, 23-24
density, 18
diagnostic features, 82 (table)
digestive upsets, 23
dry conditions, 33
following sward damage, 33
grazing by sheep, 18
herbicides for control, 23
impeding harvesting, 23
in grazing, 36
in newly sown perennial ryegrass, 14
in permanent pastures, 42
nitrogen, 8
susceptibility to herbicides, 22 (table), 38
(table)
when to spray, 38
Chlorflurecol-methyl + maleic hydrazide, cow
parsley, 42
Chlorothalonil
crown rust, 54
drechslera foot rot, 53
drechslera leaf spots, 54, 64
fusarium root rot, 53
pseudopeziza leaf spot, 74
Chlorpyrifos, 10
frit fly, 26
leatherjackets, 27, 49
Choke, pl. VI
description, 67
losses, 67
Chrysanthemum segetum — see Corn marigold

Cicuta virosa — *see* Cowbane
Cirsium acaule — *see* Dwarf thistle
Cirsium arvense, biological control, 9 — *see also* Creeping thistle
Cirsium palustre — *see* Marsh thistle
Cirsium vulgare — *see* Spear thistle
Cirsium spp., 9
Cladosporium leaf spot of timothy
  description, 65
  losses, 65
Cladosporium phlei, losses, 65
Cladochytrium spp. — *see* Damping off, post-emergence
Claviceps purpurea — *see* Ergot
Cleavers
  herbicides for control, 23
  susceptibility to herbicides, 22 (table)
Clover phyllody, pl. VII
  nodulation, 78
  symptoms, 78
  treatments, 78
Clover red leaf, pl. VII, 79
  symptoms, 79
  transmission, 79
Clover rot, pl. VII, 72-73
  cutting, 72
  description, 72
  grazing, 72
  life cycle, 72
  losses, 72
  management, 72
  nitrogen, 72
  resistance, 73
Clover scorch, pl. VII, 73
  conditions favouring, 73
  description, 73
  losses, 73
Clover seedling, 21 (fig.)
Clover stolon growth, reduction by weeds, 1
Clover weevils
  description, 51
  life cycle, 51
Clover witches' broom, symptoms, 78-79
Clover yellow vein virus, 77
  transmission, 77
Clover — *see also* Alsike clover, Red clover, White clover
Clover-safe herbicides, 9, 35, 37
Clovers
  clover red leaf, 78
  oestrogen content, 74
  phyllody, 78
  unsafe herbicides, 46
  weeds damaging to, 18
  witches' broom, 78
Cockchafer, life cycle, 50
Cocksfoot, 10

bacterial wilt of ryegrass, 67
BYDV, 70
blind seed disease of ryegrass, 66
Cambria, 71
chafers, 50
choke, 67
cocksfoot leaf fleck, 64
CfMV, 71
drechslera leaf spots, 64
frit fly, 49
halo spot, 65
identification (key), 84
losses due to choke, 67
powdery mildew, 63
rhynchosporium leaf blotch, 63
RMV, 68
snow mould, 53
stripe rust, 63
stripe smut, 65
Cocksfoot aphid, virus transmission, 51
Cocksfoot leaf fleck, 64-65
  cutting, 65
  description, 64
  grazing, 65
  losses, 64
  nitrogen, 65
  potash, 65
  weather conditions, 64
Cocksfoot mottle virus, 3, 71
  resistance, 71
  spread, 71
Cocksfoot streak virus, 51, 71
  description, 71
  transmission, 71
Colchicum autumnale — *see* Meadow saffron
Common chickweed — *see* Chickweed
Common couch — *see* Couch
Common fumitory — *see* Fumitory
Common hemp nettle — *see* Hemp nettle
Common nettle — *see* Nettle
Common ragwort — *see* Ragwort
Common rush — *see* Rush, common
Common sorrel — *see* Sorrel
Common speedwell — *see* Speedwell
Compact rush — *see* Rush, compact
Conium maculatum — *see* Hemlock
Control methods
  compatibility of techniques, 11
  summary of non-chemical, 8 (table)
Control techniques
  additive effects, 14
  herbicide + insecticide + fungicide, 14
  pesticide + fungicide, 14
Corn marigold
  herbicides for control, 23
  susceptibility to herbicides, 22 (table)
Corn spurrey

diagnostic features, 82 (table)
susceptibility to herbicides, 22 (table)
Couch, pl. I, 1
flat smut, 65
high N, 32
identification (key), 85
Couch grass — see Couch
Cow parsley
in all-grass swards, 42
resistance to herbicides, 42
Crambus spp. — see Grass moths
Cranefly, 10
Creeping buttercup — see Buttercup, creeping
Creeping soft grass, pl. II
identification (key), 85
sensitivity to asulam, 43
Creeping thistle — see Thistle, creeping
Crested dogstail
identification (key), 84, 86
snow mould, 53
Crimson clover
rust, 74
WCMV, 76
Crown rust, pl. VI, 2
chemical control, 54
cutting, 54
damage, 54
description, 54
grazing, 54
life cycle, 54
losses, 3, 54
nitrogen, 4, 54
resistance, 54
suppressed by ryegrass mosaic virus, 4
susceptibility of Lior, 4
sward imbalance, 54
weather conditions, 54
CSV — see Cocksfoot streak virus
Culm, 83 (fig.)
Cultivars
development for disease resistance, 4
disease resistant, 5
Cultivation, to reduce pest populations, 2
Cultural control, 15
diseases, 36
insect pests, 10
methods, 7, 15
pathogens, 10
weeds, 18, 37
wireworms, 28
Curled dock, occurrence, 43
Cutting
aphids, 51
bacterial wilt of ryegrass, 68
clover rot, 72
cocksfoot leaf fleck, 65
crown rust, 54

frequency, 33-34
halo spot, 65
pseudopeziza leaf spot, 74
stripe rust, 63
Cutting and grazing, alternate, 36
CYVV — see Clover yellow vein virus
Cylindrocarpon spp. — see Damping off
Cynosurus cristatus — see Crested dogstail
Cypermethrin, frit fly, 26

2,4-D, 9, 14
all-grass leys, 22
broad-leaved weeds, susceptibility, 38
(table)
effect on rust, 10
grass-only swards, 37
minimal impact on insects, 11
not for red clover, 41
ragwort, 46
— see also 2,4-DB + 2,4-D + MCPA
2,4,5-T + 2,4-D
2,4,5-T + 2,4-D ± dicamba
2,4-D + MCPA
nettles, 45
2,4-D amine
buttercups, 41
docks, 43
marsh horsetail, 44
perennial ryegrass, amount tolerated, 21
soft rush, 46
mixtures
buttercups, 41
2,4-D ester
buttercups, 41
docks, 43
marsh horsetail, 44
perennial ryegrass, amount tolerated, 21
ragwort, 46
soft rush, 46
mixtures
buttercups, 41
Dactylis glomerata — see Cocksfoot
Daisy, susceptibility to herbicides, 38 (table)
Dalapon — see TCA + dalapon
Dalapon sodium, tussock grass, 47
Damping-off, 2
cultural control, 28
post-emergence, 28
pre-emergence, 28
seed dressing, 28-29
Dandelion
in grazing, 36
susceptibility to herbicides, 38 (table)
sward defoliation, 34
Danthonia decumbens — see Heath grass
2,4-DB, 9
all-grass leys, 22

2,4-DB—*cont.*
  broad-leaved weeds, susceptibility, 38
    (table)
  buttercups, 41
  grass/clover leys, 22
  mixtures
    buttercups, 41
    resistance of ragwort, 46
  salt, perennial ryegrass, amount tolerated,
    21
  swards containing clover, 37
  thistles, 47
  — *see also* Benazolin + 2,4-DB + MCPA
    Linuron + 2,4-DB + MCPA
2,4-DB + 2,4-D + MCPA
  all-grass leys, 22
  broad-leaved weeds, susceptibility, 38
    (table)
  grass/clover leys, 22
  swards containing clovers, 37
2,4-DB + MCPA
  all-grass leys, 22
  chickweed, 24
  grass/clover leys, 22
2,4-DB ± MCPB, nettles, 45
2,4-DB and mixtures
  black bindweed, 23
  knotgrass, 23
  redshank, 23
DDT, 49
  not permitted, 28
Dead-nettle
  susceptibility to herbicides, 22 (table)
Defoliation
  frequency, 33-34
  pathogen control, 11
Demephion, aphids, 77
Demeton-S-methyl, aphids, 51, 70, 77
Deroceras reticulatus — *see* Slugs
Desiccating old pasture, frit fly, 2
Deschampsia caespitosa — *see* Tufted hair
  grass, Tussock grass
Deschampsia flexuosa — *see* Wavy hair-grass
Dicamba — *see* 2,4,5-T + 2,4-D ± dicamba
  2,3,6-TBA + dicamba + MCPA + mecoprop
  Triclopyr + dicamba + mecoprop
Dicamba + dichlorprop + MCPA
  all-grass leys, 22
Dicamba + mecoprop
  broad-leaved weeds, susceptibility, 38
    (table)
  docks, 44
  grass-only swards, 37
Dicamba + mecoprop + MCPA
  all-grass leys, 22
  broad-leaved weeds, susceptibility, 38
    (table)

docks, 44
grass-only swards, 37
Dicamba + mecoprop + 2,4,5-T
  broad-leaved weeds, susceptibility, 38
    (table)
  docks, 44
  grass-only swards, 37
  ragwort, 46
Dichlofop-methyl
  not clover-safe, 25
  wild oats, 25
3,6-dichloropicolinic acid
  broad-leaved weeds, susceptibility, 38
    (table)
  grass-only swards, 37
  thistles, 47
  — *see also* Triclopyr + 3,6-dichlorpicolinic
    acid
Dichlorprop
  cow parsley, 42
  — *see also* Dicamba + dichlorprop + MCPA
Dieldrin, not permitted, 28
Difenzoquat, wild oats, 25, 26
Digestive upsets
  chickweed, 23
  in grazing animals, 3
Digitalis purpurea — *see* Foxglove
Dilophus febrilis — *see* Fever fly
Dimethoate, aphids, 51, 70
2,5-dimethyl-3-furylanide, flag smut, 66
Dinoseb
  chickweed, 23
  corn marigold, 23
  cleavers, 23
  mayweed, 23
  acetate
    all-grass leys, 22
    chickweed, 24
    grass/clover leys, 22
  amine
    all-grass leys, 22
    chickweed, 24
    grass/clover leys, 22
    perennial ryegrass, amount tolerated, 21
Dinoseb acetate/amine
  broad-leaved weeds, susceptibility, 38
    (table)
  swards containing clover, 37
Direct drilling, pest build-up, 2
Disease ratings, grass cultivars, 5
Diseases
  bacterial in established swards, 67-68
  crown and root of herbage legumes, 72-73
  cultural control, 36
  foot and root, 53-54
  grasses, 53-71
  grassland, chemical control, 4

herbage legumes, 72-79
  in grass and clover, 2-5
  mycoplasma, symptoms, 78
  seedling and establishment, 28-29
  stem and foliage in established swards, 54,
    63-66
  stem and leaf of herbage legumes, 73-75
  vectors, 2
  virus, established swards, 68-71
    herbage legumes, 75-77
Disulfoton, aphids, 77
Ditylenchus — *see* Eelworms
Dock, broad, susceptibility to herbicides, 38
  (table)
Dock, curled, susceptibility to herbicides, 38
  (table)
Docks, 9, 11
  cutting, 35
  effect of nitrogen, 8, 32
  in all-grass swards, 43
  in grass/clover swards, 43
  in grazing, 36
  management, 43
  slurry application, 32
  susceptibility to herbicides, 22 (table)
  when to spray, 38
Downy mildew, pl. VII, 74
  description, 74
Drainage, 33
Drazoxolon, seed dressing, 29
Drechslera andersenii — *see* Drechslera leaf
  spots
Drechslera catenaria — *see* Drechslera leaf
  spots
Drechslera dictyoides — *see* Drechslera leaf
  spots
Drechslera dictyoides f.sp. *dictyoides* — *see*
  Drechslera leaf spots
Drechslera dictyoides f.sp. *perenne* — *see*
  Drechslera leaf spots
Drechslera festucae — *see* Drechslera leaf
  spots
Drechslera leaf spots, 54
  cutting, 64
  damage, 64
  description, 64
  grazing, 64
  losses, 64
  nitrogen, 4, 64
  resistance, 64
  susceptibility of Lior, 4
  sward composition, 64
Drechslera leaf streak, 64
Drechslera nobleae — *see* Drechslera leaf spots
Drechslera phlei — *see* Drechslera leaf spots
Drechslera siccans, in mixed sward, 3 — *see*
  also Drechslera leaf spots

Drechslera sorokiniana — *see* Drechslera leaf
  spots
Drechslera spp. — *see* Damping off
Drought, 33
Dung, 36
  sward problems, 34
Dwarf thistle — *see* Thistle, dwarf

Eelworms
  damage, 27
  difficulty of control, 27
  in establishing swards, 27
Elymus repens — *see* Couch
English stolbur — *see* Clover red leaf
Epichloe typhina — *see* Choke
Equisetum arvense — *see* Horsetail
Equisetum palustre — *see* Marsh horsetail
Ergot, 66-67
  description, 66
  disorders in livestock, 66
  management, 67
Eriophyid mites, 2
Erysiphe graminis — *see* Powdery mildew
Erysiphe trifolii — *see* Powdery mildew
Established swards
  broad-leaved weed control, 37 (table)
  weeds, occurrence and control, 39-47
Establishment, conditions for success, 17
Ethofumesate
  annual grass weeds, 19
  annual meadow grass in all-grass leys, 25
  and seed rate, 14
  annual weed grasses, 23
  barley grass, 39
  chickweed, 19
  pre-emergence spraying, 19
  soft brome, 42
  unsuitable for clover, 19
  volunteer cereals, 19, 26
Ethofumesate + phorate + benomyl, 14
Euscelis plebeja
  clover phyllody transmission, 78
  clover red leaf transmission, 79

False oat-grass, pl. III
  identification (key), 85
Fat hen, 18
  diagnostic features, 82 (table)
  susceptibility to herbicides, 22 (table)
Fertilizers
  amounts, 17
  effect on establishment, 17
Fescues
  blind seed disease of ryegrass, 66
  BYDV, 70
  choke, 67

Fescues—*cont.*
crown rust, 54
drechslera leaf spots, 64
— *see also* Meadow fescue, Red fescue, Sheep's fescue, Tall fescue
Festuca arundinacea — *see* Tall fescue
Festuca ovina — *see* Sheep's fescue
Festuca pratensis — *see* Meadow fescue
Festuca rubra — *see* Red fescue
Fever fly, 51
Field horsetail — *see* Horsetail
Field pansy — *see* Pansy
Field penny-cress — *see* Penny-cress
Field poppy — *see* Poppy
Flamprop-methyl, wild oats, 25
Flat smut, 65
Fog, crown rust, 54
Foot and root diseases — *see* Diseases
Foxglove, 39
Frit fly, pl. V, 2, 10
control in establishing swards, 26
damage, 26
damage threshold, 49
in establishing swards, 26
in established grasses, 49
life cycle, 26
recognition, 26
Fumaria officinalis — *see* Fumitory
diagnostic features, 82 (table)
susceptibility to herbicides, 22 (table)
Fungi, for weed control 10
Fungus diseases — *see* Diseases
Fusarium avenaceum — *see* Fusarium root rot complex
Fusarium culmorum — *see* Fusarium root rot
Fusarium moniliforme — *see* Fusarium rot rot complex
Fusarium nivale — *see* Fusarium patch, snow mould
Fusarium oxysporum — *see* Fusarium root rot complex
Fusarium patch, in established swards, 53
Fusarium root rot
conditions leading to, 73
in established swards, 53
management, 73
potash, 73
Fusarium root rot complex, 73
Fusarium roseum — *see* Fusarium root rot complex
Fusarium solani — *see* Fusarium root rot complex
Fusarium spp. — *see* Damping off

Gaeumannomyces graminis — *see* Take-all
Galeopsis spp. — *see* Hemp nettle
Galium aparine — *see* Cleavers

Gamma-HCH
chafers, 50
leatherjackets, 49
and bran leatherjackets, 27
Garden chafer, life cycle, 49-50
Gastrophysa viridula, attacking docks, 11
Geomyza tripunctata — *see* Frit fly
Germination, effect on establishment, 17
Ghost swift moth — *see* Swift moth
Gibberellic acid, choke, 67
Gloeotinia temulenta — *see* Blind seed disease of ryegrass
Glyphosate
bracken, 40
docks, 43
nettles, 45
ragwort, 46
soft rush, 46
thistles, 47
Grain aphid, BYDV transmission, 69
Grass, diagnostic characters, 83 (fig.)
Grass/clover leys, newly sown
broad leaved weed control, 22 (table)
weed susceptibility to herbicides, 22 (table)
Grass cultivars, disease ratings, 5
Grass moths, 51-52
description, larva and adult, 51
in grassland, 52
life cycle, 51
management, 52
Grass, seedling growth stages, 20 (fig.)
Grass tillering, reduction by weeds, 1
Grasses, identification, 83-87
(key), 84-86
Grassland
composition and pH, 31
invasion by weeds, 1
Grassland diseases
chemical control, 4
management practices, 4
Grazing
aphids, 51
chickweed, 18
clover rot, 72
cocksfoot leaf fleck, 65
crown rust, 54
drechslera leaf spots, 64
halo spot, 65
ryegrass mosaic virus, 69
seasonality, 34
stripe rust, 63
Grazing and cutting, alternate, 36
Green petal disease, 78
Groundsel
diagnostic features, 82 (table)
susceptibility to herbicides, 22 (table)

Halo spot
  cutting, 65
  description, 65
  grazing, 65
  nitrogen, 65
Hard rush — see Rush, hard
Hawkbit, susceptibility to herbicides, 38 (table)
Heath grass, identification (key), 84
Heath rush — see Rush, heath
Hemlock, 39
Hemp nettle
  herbicide for control, 23
  susceptibility to herbicides, 22 (table)
Hepialus humuli — see Ghost swift moth
Herbage legumes
  disease resistance, 5
  diseases, 72-79
Herbicides, 9
  application, 38-39
  choice of treatment, 37
  clover-safe 9, 37
  for specific weeds, newly sown leys, 23 (table)
  management after spraying, 39
  resistance of cow parsley, 42
  resistance of ragwort, 46
  safety, 39
  susceptibility of broad-leaved weeds, 22 (table)
  susceptibility of broad-leaved weed seedlings, 22 (table)
  time before grazing, 39
  unsafe for clovers, 46
  weather conditions, 38
  when to spray, 38
Heterosporium, 65
Holcus lanatus — see Yorkshire fog
Holcus mollis — see Creeping soft grass
Hordeum murinum — see Barley grass, Wall barley
Hordeum secalinum — see Meadow barley
Hordeum spp. — see Barley
Horses
  bracken poisoning, 40
  ragwort poisoning, 45
  selectivity, 34
Horsetail, 1, 39
  — see also Marsh horsetail
Horsetails, susceptibility to herbicides, 38 (table)
Hyalopteroides humilis — see Cocksfoot aphid

Inflorescence diseases, established swards, 66-67
Injury to livestock, 1
Insecticides, 10

Integrated control, definition, 7
Integrated weed management, 9
Interactions
  control techniques , 11, 12, 13
  weeds and other pests, 12, 14
Ioxynil — see Bromoxynil + ioxynil, Bromoxynil + ioxynil + mecoprop
Irrigation, 33
  white clover, 35
Italian/perennial ryegrass
  bacterial wilt of ryegrass, 67
Italian ryegrass
  bacterial wilt of ryegrass, 67
  blind seed disease of ryegrass, 66
  crown rust, 54
  fusarium root rot, 53
  identification (key), 85
  Ligniera junci, 53
  powdery mildew resistance, 63
  resistance to RMV, 69
  rhynchosporium leaf blotch, 63
  RMV, 68-69
  snow mould, 53
  take-all, 53
Ivy-leaved speedwell — see Speedwell

Japanese beetle, 10
Jointed rush — see Rush, jointed
Juncus articulatus — see Rush, jointed
Juncus effusus — see Rush, common, Rush, soft
Juncus inflexus — see Rush, hard
Juncus squarrosus — see Rush, heath
Juncus spp. — see Rushes

K — see Potash
Kabatiella caulivora — see Clover scorch
Knapweed, susceptibility to herbicides, 38 (table)
Knot grass
  diagnostic features, 82 (table)
  herbicides for control, 23
  susceptibility to herbicides, 22 (table)

L-flamprop-isopropyl, wild oats, 25
Leafhoppers, 2
  mycoplasma transmission, 78
Leatherjackets, pl. V, 2
  damage, 27
  description, 27
  in established grassland, 49
  in establishing swards, 27
  life cycle, 27
  monitoring and forecasting, 27
Leptosphaerulina trifolii — see Burn
Leys, newly sown, herbicides for specific weed control, 22 (table)

Ligniera junci
  in established swards, 53
  in mixed sward, 3
Ligule, 83 (fig.)
Liming, 31
Linuron
  all-grass leys, 22
  chickweed, 24
Linuron +2,4-DB + MCPA
  all grass leys, 22
  chickweed, 24
  grass/clover leys, 22
Lior, disease susceptibility, 4
Lolium multiflorum — *see* Italian ryegrass
Lolium perenne — *see* Perennial ryegrass
Loose smut, description, 66
Losses
  barley yellow dwarf virus, 70
  blind seed disease of ryegrass, 66
  BYMV and AMV, 76
  choke, 67
  Cladosporium phlei, 65
  clover rot, 72
  clover scorch, 73
  cocksfoot leaf fleck, 64
  crown rust, 3, 54
  drechslera leaf spot, 64
  due to disease, 3
  due to pests, 2
  powdery mildew, 3, 63
  rhynchosporium leaf blotch, 64
  RMV, 69
  WCMV, 76
Lucerne
  AMV, 76
  fusarium root rot complex, 73
  pseudopeziza leaf spot, 74
  WCMV, 76
Lucerne mosaic virus — *see* Alfalfa mosaic virus

Macrosiphum euphorbiae — *see* Potato aphid
Macrosiphum pisi — *see* Pea aphid
Macrosteles sexnotatus, mycoplasma transmission, 79
Macrosteles spp., clover phyllody transmission, 78
Maleic hydrazide — *see* Chlorflurecol-methyl + maleic hydrazide
Management
  grassland diseases, 4
  pest and weed control, 7
Maneb — *see* Nickel sulphate + maneb
Manure
  farmyard, 32
  sward damage, 32-33
Marigold, corn

susceptibility to herbicides, 22 (table)
Marsh horsetail
  management, 44
  poisoning by, 44
Marsh ragwort, poisonousness, 45
Marsh thistle — *see* Thistle, marsh
Mastigosporium album, 64
Mastigosporium cylindricum, 64
Mastigosporium rubricosum — *see* Cocksfoot leaf fleck
Mat grass
  identification (key), 84
  pH, 31
Matricaria perforata — *see* Mayweed, scentless
Matricaria spp. — *see* Mayweeds
Mayweed, scentless
  diagnostic features, 82 (table)
  susceptibility to herbicides, 22 (table)
Mayweeds, herbicides for control, 23
MCPA, 9
  all-grass leys, 22
  broad-leaved weeds, susceptibility, 38 (table)
  docks, 43
  grass only swards, 37
  marsh horsetail, 44
  ragwort, 46
  soft rush, 46
  thistles, 47
  — *see also* Asulam + mecoprop + MCPA
  Benazolin + 2,4-DB + MCPA
  Bentazone + MCPB + MCPA
  2,4-D + MCPA
  2,4-DB + MCPA
  2,4-DB + 2,4-D + MCPA
  Dicamba + dichlorprop + MCPA
  Dicamba + mecoprop + MCPA
  Linuron + 2,4-DB + MCPA
  MCPB + MCPA
  2,3,6-TBA + dicamba + MCPA + mecoprop
MCPA mixtures, buttercups, 41
MCPA salt
  buttercups, 41
  perennial ryegrass, amount tolerated, 21
MCPB, 9
  all-grass leys, 22
  broad-leaved weeds, susceptibility, 38 (table)
  buttercups, 41
  grass/clover leys, 22
  resistance of ragwort, 46
  swards containing clover, 37
  thistles, 47
  mixtures
    buttercups, 41

hemp nettle, 23
salt
  perennial ryegrass, amount tolerated, 21
  — *see also* Bentazone + MCPB + MCPA
    2,4-DB ± MCPB
MCPB + MCPA
  all-grass leys, 22
  broad-leaved weeds, susceptibility, 38
    (table)
  creeping buttercups, 41
  grass/clover leys, 22
  swards containing clover, 37
Meadow barley
  identification (key), 85
Meadow buttercup — *see* Buttercup, meadow
Meadow fescue, 10
  bacterial wilt of ryegrass, 67
  identification (key), 85
  RMV, 68
Meadow foxtail, identification (key), 86
Meadow grasses, 19
  blind seed disease of ryegrass, 66
  choke, 67
  control in perennial ryegrass, 25
  flat smut, 65
  sensitivity to asulam, 43
Meadow saffron, 39
Meadowsweet
  susceptibility to herbicides, 38 (table)
Mechanical control, 7, 9
  insect pests, 10
  methods, 8 (table)
Mecoprop
  all-grass leys, 22
  broad-leaved weeds, susceptibility, 38
    (table)
  chickweed, 14, 24
  docks, 44
  grass-only swards, 37
  nettles, 45
  thistles, 47
  — *see also* Asulam + mecoprop + MCPA
    Dicamba + mecoprop
    Dicamba + mecoprop + MCPA
    Dicamba + mecoprop + 2,4,5-T
    2,3,6-TBA + dicamba + MCPA + meco-
      prop
    Triclopyr + dicamba + mecoprop
Mecoprop salt, perennial ryegrass, amount
  tolerated, 21
Mecoprop + asulam + MCPA, nettles, 45
Mefluidide, barley grass, 39
Megoura viciae — *see* Vetch aphid
Meloidogyne naasi — *see* Eelworms
Melolontha melolontha — *see* Cockchafer
Metaldehyde, slugs, 27, 49
Methabenzthiazuron

black grass, 19
chickweed, 19
crop under stress, 19
meadow grasses, 19, 25
not clover-safe, 19
pre-emergence spraying, 19
undersown crops, 19
Methiocarb, slugs, 27, 49
Metopolophium dirhodum — *see* Rose-grain
  aphid
Metopolophium festucae, 50
Micronectriella nivalis
  — *see* Fusarium patch, Snow mould
Mildew, susceptibility of Lior, 4
Milky disease, control of Japanese beetle, 10
Minimum tillage methods, pest build-up, 2
Mites, 2
  breeding for resistance to, 69
  RMV, 69
MLO — *see* Mycoplasma
Molinia caerulea — *see* Purple moor grass
Monoculture, risk of disease, 3
Moth, bracken control, 9
Mowing, weeds, 18
Musk thistle, 46
Mycoplasma
  diseases, symptoms, 78
  transmission, 78
Myzus persicae — *see* Peach-potato aphid

N — *see* Nitrogen
Nardus stricta — *see* Bent wire grass, Mat
  grass
Nematodes, 2 — *see also* Eelworms
Neodusmetia sangwani, 10
Net blotch, 64
Nettle, susceptibility to herbicides, 38 (table)
Nettles
  in all-grass swards, 45
  in grass/clover swards, 45
  in grazing, 36
  management, 45
  small, susceptibility to herbicides, 22 (table)
  when to spray, 38
Newly sown fields, pest damage, 2
Nickel sulphate + maneb
  crown rust, 54
  stripe rust, 63
Nitrogen
  bacterial wilt of ryegrass, 68
  clover rot, 72
  cocksfoot leaf fleck, 65
  crown rust, 54
  deficiency in soil, 32
  drechslera leaf spot, 64
  effect on fungus diseases, 4-5
  effect on weeds, 8

Nitrogen—*cont.*
 fertilizer, 32
 halo spot, 65
 powdery mildew, 63
 RMV, 69
 stripe rust, 63
 sward diversity, 32
 white clover, 35
Northern anthracnose, 73
Nodulation, clover phyllody, 78
Nutritive quality, fungus infected grassland, 3

Oats, pl. IV
 identification (key), 86
 RMV, 68
Oenanthe spp. — *see* Water-dropworts
Oestrogen
 burn, 75
 herbage legumes, 74
 pseudopeziza leaf spot, 74
 rusts, 74
Old swards, pest build-up, 2
Omethoate, frit fly, 26
Organo-mercury
 clover scorch, 73
 damping off, 29
Oscinella frit — *see* Frit fly
Oulema lichenis, CMV, 71
Oulema melanopus — *see* Cereal leaf beetle
Oxycarboxin
 crown rust, 54
 flag smut, 66
Oxydemeton methyl, aphids, 51, 70, 77

P — *see* Phosphate
Pansy, field, susceptibility to herbicides, 22 (table)
Paraquat
 cereal stubbles, 19
 pre-emergence spraying, 19
 seedling weeds, 19
Parthenodes angularis, 9
Parsley piert, susceptibility to herbicides, 22 (table)
Pea aphid
 AMV transmission, 76
 BYMV transmission, 77
 CYVV transmission, 77
 WCMV transmission, 76
Pea mosaic, 76
Peas
 CYVV, 77
 RCVMV, 77
 WCMV, 76
Peach-potato aphid
 AMV transmission, 76

BYMV transmission, 77
CSV transmission, 71
CYVV transmission, 77
Penny cress, field, susceptibility to herbicides, 22 (table)
Pepper spot — *see* Burn
Perennial ryegrass
 BYDV, 70
 blind seed disease of ryegrass, 66
 cocksfoot leaf fleck, 64
 crown rust, 54
 fusarium root rot, 53
 grass moths, 52
 herbicide tolerance at 3-leaf, 21 (table)
 high N, 32
 identification (key), 84
 Ligniera junci, 53
 Mascot, 69
 Mastigosporium album, 64
 pre-emergence weed control, 19
 resistance to RMV, 69
 seed depth, 18
 snow mould, 53
 soil fertility, 31
 treading, 34
Perennial weeds, 18
Peronospora trifoliorum — *see* Downy mildew
Pesticide treatments, control of establishment pests, 2
Pests
 established grasses and legumes, 49-52
 establishing swards, 26-28
 in grass and clover, 1-2
Phleum pratense — *see* Timothy
Phorate
 aphids, 77
Phorate + ethofumesate + benomyl, 14
Phosphate, requirements, 32
Phyllody — *see* Clover phyllody
Phyllopertha horticola, attacking bracken, 11
 — *see also* Garden chafer
Pirimicarb, aphids, 51, 70, 77
Plant-hoppers, 2
Plantago spp. — *see* Plantains
Plantains, susceptibility to herbicides, 38 (table)
 sward defoliation, 34
Poa annua — *see* Annual meadow grass
Poa pratensis — *see* Smooth meadow grass
Poa trivialis, high N, 32
 — *see also* Rough meadow grass
Poaching, 34
 grazing and cutting frequency, 34
 wet soils, 33
Poisoning
 horsetail, 44
 ragwort, 35, 45

Poisonous plants, 1
Poisonous weeds, 39
Polygonum aviculare — *see* Knotgrass
Polygonum convolvulus — *see* Black bindweed
Polygonum persicaria — *see* Redshank
Popillia japonica — *see* Japanese beetle
Poppy, susceptibility to herbicides, 22 (table)
Post-emergence spraying
  crop growth stage, 19
  pre-treatment mowing, 21
  weed growth stage, 21
Potash
  cocksfoot leaf fleck, 65
  fusarium root rot, 73
  in cut fields, 36
  requirements, 32
Potato aphid
  BYMV transmission, 77
  CYVV transmission, 77
Potatoes, stolbur, 79
Powdery mildew, 73-74
  description, 63
  losses, 3, 63
  management, 63
  nitrogen, 63
  resistance, 63
  weather conditions, 63
Pratylenchus — *see* Eelworms
Pseudopeziza leaf spot of clovers
  cutting, 74
  description, 74
  oestrogen, 74
Pseudopeziza trifolii — *see* Pseudopeziza leaf spot of clovers
Pteridium aquilinum — *see* Bracken
Pteridium esculentum — *see* Bracken
Puccinia coronata — *see* Crown rust
Puccinia coronata f.sp. *festucae* — *see* Crown rust
Puccinia coronata f.sp. *lolii* — *see* Crown rust
Puccinia recondita f.sp. *lolii* — *see* Brown rust
Puccinia striiformis — *see* Stripe rust
Purple moor grass, pH, 31
Pythium spp. — *see* Damping off, Fusarium root rot complex

Quintozene
  clover rot, 72
  snow mould, 54

Radish, wild, susceptibility to herbicides, 22 (table)
Ragwort, 1, 11
  common, appearance, 45
  cutting, 35, 37, 46
  grazing by sheep, 46
  in grazing, 36

in hay, 39
marsh, appearance, 45
poison risk, 35
poisonousness, 45
poisoning symptoms, 45
precautions after spraying, 45
pulling or digging, 46
resistance to herbicides, 46
selective grazing, 34
susceptibility to herbicides, 38 (table)
timing of spraying, 46
when to spray, 38
— *see also* Marsh ragwort
Ranunculus acris — *see* Meadow buttercup
Ranunculus bulbosus — *see* Bulbous buttercup
Ranunculus repens — *see* Creeping buttercup
Ranunculus spp. — *see* Buttercups
RCNMV — *see* Red clover necrotic mosaic virus
RCSV — *see* Ryegrass chlorotic streak virus
RCVMV — *see* Red clover vein mosaic virus
Red clover
  AMV, 76
  BYMV, 76
  clover phyllody, 78
  clover red leaf, 79
  clover rot, 72
  clover scorch, 73
  CYVV, 77
  downy mildew, 74
  fusarium root rot complex, 73
  Hungaropoly, 76
  Lakeland, 73
  powdery mildew, 73
  pseudopeziza leaf spot, 74
  RCNMV, 76
  RCVMV, 77
  rust, 74
  WCMV, 76
Red fescue
  bacterial wilt of ryegrass, 67
  identification (key), 84
  P and K deficiency, 32
  snow mould, 53
  stripe smut, 65
Red clover necrotic mosaic virus, pl. VII, 75
  transmission, 76
Red clover vein mosaic virus
  symptoms, 77
  transmission, 77
Redshank
  herbicides for control, 23
  susceptibility to herbicides, 22 (table)
Reproduction, animal
  effects of diseased herbage, 8
  pseudopeziza leaf spot, 74

Reproductive disorders, animal, herbage
  legume oestrogen content, 74
Re-seeds, pest risk, 2
Resistance
  clover rot, 73
  clover scorch, 73
  cultivars, 5
  rhynchosporium leaf blotch, 64
Rhamnus sp. — *see* Buckthorn
Rhinocyllus conicus, 9
Rhizoctonia solani — *see* Damping off, post-
  emergence
Rhizoctonia spp. — see Fusarium root rot
  complex
Rhizopus spp. — *see* Fusarium root rot complex
Rhodes grass scale, 10
Rhopalosiphum padi, 50
  RCSV transmission, 70
  — *see also* Bird-cherry aphid
Rhynchosporium leaf blotch
  description, 63-64
  losses, 3, 64
  resistance, 64
  susceptibility of Lior, 4
Rhynchosporium orthosporum — *see* Rhyn-
  chosporium leaf blotch
Rhynchosporium secalis — *see* Rhynchospor-
  ium leaf blotch
RMV — *see* Ryegrass mosaic virus
Rose-grain aphid
  BYDV transmission, 69
Rough meadow grass, pl. I
  identification (key), 84
  in newly-sown leys, 25
  — *see also* Meadow grasses
Rumex acetosa — *see* Sorrel
Rumex crispus — *see* Curled dock
Rumex obtusifolius — *see* Broad-leaved dock
Rumex spp. — *see* Docks
Rush
  common, cutting, 35
  compact, susceptibility to herbicides, 38
    (table)
  hard, 46
    susceptibility to herbicides, 38 (table)
  heath, susceptibility to herbicides, 38
    (table)
  jointed, 46
  soft
    distinguishing, 46
    susceptibility to herbicides, 38 (table)
Rushes, 1, 18
  management, 46
  prevention, 8
  water-logged soils, 33
Rust

effect of 2,4-D, 10
  herbage legumes, description, 74-75
Rusted herbage, rejection by cattle, 3
Rusts, 74-75
  herbage legume oestrogen content, 74
Ryegrasses
  bacterial wilt of ryegrass, 67
  BYDV, 70
  brown rust, 54
  choke, 67
  crown rust, 54
  crown rust resistance, 54
  drechslera leaf spots, 64
  drechslera leaf spot resistance, 64
  flat smut, 65
  frit fly, 49
  powdery mildew, 63
  pre-emergence weed control, 19
  resistance to bacterial wilt, 68
  rhynchosporium leaf blotch resistance, 64
  RMV, 68
  RCSV, 70
  snow mould resistance, 54
  stripe smut, 65
Ryegrass chlorotic streak virus
  description, 70
  transmission, 70
Ryegrass mosaic virus, pl. VI, 68-69
  description, 68
  effect on crown rust, 4
  grazing, 69
  losses, 3, 69
  management, 69
  nitrogen, 5, 69
  resistance, 69
  slot seeding, 69
  time of sowing, 5
  transmission, 69
  with BYDV, 4

Sainfoin, powdery mildew, 73
St. Mark's fly, 51
Safety
  precautions, 88
  regulations, 88
Sclerotinia trifoliorum — *see* Clover rot
Seed depth, perennial ryegrass, 18
Seed dressings, 4, 10
Seed rate, effect on establishment, 17
Seed treatments
  clover scorch, 73
  control of establishment pests, 2
Seedling weeds, paraquat, 19
Selenophoma donacis — *see* Halo spot
Selfheal, susceptibility to herbicides, 38,
  (table)
Senecio aquaticus — *see* Marsh ragwort

Senecio jacobaea — *see* Ragwort
Senecio vulgaris — *see* Groundsel
Sheath, 83 (fig.)
Sheep
  bracken poisoning, 40
  ergot disorders, 66
  ragwort poisoning, 45
  selectivity, 34
  tick-borne diseases, 40
  winter grazing, 34
Sheep's fescue, identification (key), 84
Shepherding, restricted by bracken, 1
Shepherd's purse
  diagnostic features, 82
  susceptibility to herbicides, 22 (table)
Silage, contamination, 1
Sinapis arvensis — *see* Charlock
Sitobion avenae — *see* Grain aphid
Sitobion fragariae — *see* Blackberry aphid
Sitona lineatus — *see* Clover weevils
Slot seeding
  BYDV, 70
  RMV, 69
Slugs, pl. V, 2, 10
  in established grassland, 49
  in establishing swards, 27
  life cycle, 27
  restricting movement, 27
Slurry, 32
Smooth meadow grass, identification (key), 84
Smooth sow-thistle — *see* Sow-thistle
Smooth stalked meadow grass, snow mould, 53
Smuts, 65-66
  symptoms, 65
Snow mould
  damage caused, 53
  in established swards, 53-54
  management, 53-54
  resistant cultivars, 54
Soft brome
  identification (key), 85
  management, 42
Soft rush — *see* Rush, soft
Soil
  pH, 31
  physical state, effect on establishment, 17
Sorrel,
  pH indicator, 31
  common, susceptibility to herbicides, 38 (table)
  sheep's, susceptibility to herbicides, 38 (table)
Sow-thistle
  susceptibility to herbicides, 38 (table)
  smooth, susceptibility to herbicides, 22 (table)

Spear thistle — *see* Thistle, spear
Speedwell,
  common, susceptibility to herbicides, 22 (table)
  common field, diagnostic features, 82 (table)
  ivy leaved, susceptibility to herbicides, 22 (table)
Spergula arvensis — *see* Corn spurrey
Spikelet, 83 (fig.)
Spraying
  management after, 39
  post-emergence, 19
  pre-emergence, 19
Spring burn, 63
Spurrey, corn — *see* Corn spurrey
Stellaria media — *see* Chickweed
Stocking
  mixed or alternate, 34
  mixed, 35
  rates, 35
Stolbur — *see* Clover red leaf
Stolon, 83 (fig.)
Stripe rust
  cocksfoot, 63
  cutting, 63
  description, 63
  grazing, 63
  nitrogen, 63
Stripe smut, description, 65
Sward composition, drechslera leaf spots, 64
Sward damage
  dung, 34
  manure application, 32-33
  urine, 34
  wet conditions
    livestock, 33
    machinery, 33
Sward deterioration, avoidance, 10
Sweet vernal grass, identification (key), 85
Swift moth
  description, larva and adult, 52
  life cycle, 52
  management, 52

2,4,5-T — *see* Dicamba + mecoprop + 2,4,5-T
2,4,5-T + 2,4-D
  nettles, 45
2,4,5-T + 2,4-D ± dicamba
  broad-leaved weeds, susceptibility, 38 (table)
  grass-only swards, 37
  nettles, 45
Taint, in milk, 1
Take-all
  description, 53
  in established swards, 53
  in monoculture, 3

Tall fescue
  bacterial wilt of ryegrass, 67
  identification (key), 85
Tall fescue leys, weed control, 19
Tall oat, crown rust, 54
Tall oat grass
  flat smut, 65
  loose smut, 65
Taraxacum officinale — *see* Dandelion
2,3,6 TBA + dicamba + MCPA + mecoprop
  all-grass leys, 22
  broad leaved weeds, susceptibility, 38
    (table)
  grass-only swards, 37
TCA
  volunteer cereals in perennial ryegrass, 26
TCA + dalapon
  soft-brome, 42
  barley grass, 40
Thiabendazole, flag smut, 66
Thiometon, aphids, 51, 70, 77
Thistle
  creeping, 46, 47
    cutting, 35, 37
    in grazing, 36
    rust disease, 10
    selective grazing, 34
    susceptibility to herbicides, 22 (table), 38
      (table)
  dwarf, 46
  marsh, 46
  spear, 46, 47
    susceptibility to herbicides, 38 (table)
Thistles
  integrated control, 13 (fig.)
  occurrence, 46-47
  phytophagous insects and synchronized
    spraying, 14
  prevention, 8
Tiller density, effect of weeds, 18
Time of sowing
  frit fly damage, 2
  RMV, 5
Timothy, pl. III, 10
  bacterial wilt of ryegrass, 67
  blind seed disease of ryegrass, 66
  BYDV, 70
  choke, 67
  cocksfoot leaf fleck, 64
  drechslera leaf streak, 64
  flat smut, 65
  halo spot, 65
  identification (key), 86
  Mastigosporium cylindricum, 64
  rhynchosporium leaf blotch, 63
  snow mould, 53
  stripe smut, 65

Tipula paludosa — *see* Cranefly
Tipula spp. — *see* Leatherjackets
Tobacco, stolbur, 79
Tomatoes, stolbur, 79
Topping, 34-35
Triadimefon
  crown rust, 54
  powdery mildew, 63
Triazophos, leatherjackets, 49
Trichoderma spp. — *see* Fusarium root rot
  complex
Triclopyr
  broad leaved weeds, susceptibility, 38
    (table)
  docks, 44
  grass-only swards, 37
  nettles, 45
Triclopyr + dicamba + mecoprop
  broad-leaved weeds, susceptibility, 38
    (table)
  docks, 44
  grass only swards, 37
Triclopyr + 3,6-dichloropicolinic acid
  broad-leaved weeds, susceptibility, 38
    (table)
  docks, 44
  grass-only swards, 37
Tridemorph, powdery mildew, 63, 74
Trifolium repens — *see* White clover
Triticum spp. — *see* Wheat
Tufted hair grass, 1
  identification (key), 86
Tussock grass
  in grass/clover swards, 47
  management, 47
Tylenchorhynchus — *see* Eelworms
Typhula incarnata — *see* Snow mould
Tyria jacobaeae, attacking ragwort, 11

Undersown crops
  broad-leaved weed control, 22
  weed control, 19
Urine, 36
  scorch, 34
  sward problems, 34
Urocystis agropyri — *see* Flat smut
Urocystis striiformis — *see* Stripe smut
Uromyces fallens — *see* Rusts, herbage
  legumes
Uromyces nerviphilus — *see* Rusts, herbage
  legumes
Uromyces trifolii-repentis — *see* Rusts, her-
  bage legumes
Uromyces spp. — *see* Rusts
Urtica dioica — *see* Nettle
Urtica spp. — *see* Nettles
Ustilago avenae — *see* Loose smut

Vectors, of diseases, 2
Veronica persica — see Speedwell, common
   field
Vetch aphid, 77
Virus diseases, established swards, 68-71
Viruses
   control in herbage legumes, 77
   effects on plants, 3
   multiple infection in clover, 75
   transmission by aphids, 51
   transmission, 68
   vectors, 2
   — see also individual entries
Volunteer cereals, control in newly-sown leys,
   26

Wall barley, pl. I, 1
   dry conditions, 33
   identification (key), 85
   — see also Barley grass
Wasp, parasitic, control of Rhodes grass scale,
   10
Water
   excess, 33
   shortage, 33
Water-logged soils
   damage, 33
   drainage, 33
Water stress, bacterial wilt of ryegrass, 68
Water dropworts, 39
Wavy hair-grass, identification (key), 84
WCMV— see White clover mosaic virus
Weed management, 9
Weeds
   control during establishment, 18-26
   control in permanent swards, 37-47
   in grass and clover, 1
   interaction with other pests, 12, 14
Weeds, broad-leaved
   control in established swards, 37 (table)
   control in newly sown leys, 21-24
   diagnostic features, 82 (table)
Weeds, grass, control in newly-sown leys, 25
Weevil for biological control, 9
Welted thistle, 46
Wheat
   CMV, 71
   identification (key), 85
White clover
   AMV, 76
   burn, 75
   BYMV, 76
   clover phyllody, 78
   clover red leaf, 79
   clover rot, 72
   clover scorch, 73
   clover witches' broom, 79
   competition with grass, 35

CYVV, 77
downy mildew, 74
fusarium root rot complex, 73
irrigation, 35
Ladino, 76
May grazing, 34
Menna, 76
Milkanova, 76
multiple virus infection, 75
nitrogen, 35
Olwen, 76
potential, 35
powdery mildew, 73
pseudopeziza leaf spot, 74
RCVMV, 77
rust, 74
S.100, 76
S.184, 76
safe herbicides, 41
soil conditions, 35
soil fertility, 31
viruses isolated, 75
WCMV, 76
White clover mosaic virus, 76
   losses, 76
   occurring with CYVV, 77
   symptoms, 76
Whiteheads, 67
Wild onion, 1
   susceptibility to herbicides, 38 (table)
Wild oats
   control in ryegrass seed crops, 26
   herbicides in cereals undersown, 25 (table)
Wild radish — see Radish
Winter burn, 64
Wire worms, pl. V
   chemical control not permitted, 28
   cultural control, 28
   damage, 28
   distribution, 28
   in established grass, 49
   in establishing swards, 28
   life cycle, 28
Witches' broom — see Clover witches' broom

Xanthomonas campestris pv. graminis — see
   Bacterial wilt of ryegrass
Xiphenema — see Eelworms

Yarrow, susceptibility to herbicides, 38 (table)
Yellow rattle, susceptibility to herbicides, 38
   (table)
Yorkshire fog, pl. II
   identification (key), 85
   low N, 32
   pH, 31
   sensitivity to asulam, 43

Zinc/maneb, pseudopeziza leaf spot, 74